nook
by Barnes & Noble

The Official Guide

The Official Guide

Damon Brown and Alexander Bevier

STERLING
New York

STERLING
New York

An Imprint of Sterling Publishing
387 Park Avenue South
New York, NY 10016

© 2012 by Damon Brown and Alexander Bevier

ISBN 978-1-4027-9807-8 (print format)
ISBN 978-1-4351-3883-4 (ebook)

Distributed in Canada by Sterling Publishing
c/o Canadian Manda Group, 165 Dufferin Street
Toronto, Ontario, Canada M6K 3H6
Distributed in the United Kingdom by GMC Distribution Services
Castle Place, 166 High Street, Lewes, East Sussex, England BN7 1XU
Distributed in Australia by Capricorn Link (Australia) Pty. Ltd.
P.O. Box 704, Windsor, NSW 2756, Australia

For information about custom editions, special sales, and premium and corporate purchases, please
contact Sterling Special Sales at 800-805-5489 or specialsales@sterlingpublishing.com.

Manufactured in the United States of America

10 9 8 7 6 5 4 3 2 1

www.sterlingpublishing.com

CONTENTS

NOOK Simple Touch

Appendices

Index

Introduction

Launched by Barnes & Noble in 2009, NOOK® has radically changed how we read, how we work, how we unwind, and how we travel. Instead of carrying a stack of books, we can just slide the sleek device into a carry-on and enjoy our favorite titles on the flight. New magazines can be downloaded on the fly and read on the same day they hit the newsstands. And, on the advanced model, cool apps will help you manage your finances, learn a foreign language, enjoy new music—or just play Ms. Pac-Man.

Although NOOK is commonly thought of as "just an eReader," the following pages will show you the many different things you can do with your tablet. NOOK is an excellent eReader, and you'll learn how to enjoy books, magazines, and newspapers early on in this guide. But there's much more to your NOOK than that.

In this book, we'll show you how to unlock the power in your NOOK. You'll learn how to:

- Download books
- Read periodicals
- Manage your NOOK's memory
- Troubleshoot any problems
- Choose the best accessories for you

If you have NOOK Color™, you'll also learn how to:

- Download apps
- Play games
- Listen to music
- Watch videos
- Surf the Web

And if you have NOOK Tablet™, you'll be able to:

⟩ Play HD games

⟩ Watch high-resolution movies through Netflix®

⟩ View TV shows with Hulu Plus™

⟩ Listen to music

Three NOOKs

No matter what your reading or lifestyle preferences are, NOOK has a reading device to perfectly suit your needs. NOOK Simple Touch™ (referred to in this book as simply "NOOK"), NOOK Color, and NOOK Tablet.

So, how do you choose? It is a battle of the benefits: lightweight versus graphic power, price versus apps, and simplicity versus adaptability.

NOOK Tablet	NOOK Color	NOOK Simple Touch
Weight: 14.1 oz.	Weight: 15.8 oz.	Weight: 7.48 oz.
Height: 8.1 in.	Height: 8.1 in.	Height: 6.5 in.
Width: 5.0 in.	Width: 5 in.	Width: 5 in.
Depth: 0.48 in.	Depth: .48 in.	Depth: .47 in.

NOOK Simple Touch is tailored for reading. Its black and white touchscreen E Ink® display technology makes the pages of a NOOK Book™ look just like the pages of a paper book. So if you read primarily books with black print on white paper, the NOOK may be the best device for you.

NOOK Color is in the middle ground between eReader and tablet. It can play Flash® video; run apps, including video and word games;

and use interactive, pop-up style books. The memory can also be expanded so you can download as many apps, music, videos, and reading material as you can handle.

And NOOK Tablet is a power-packed portable device, with more memory than the other two models. The app selection is formidable, too, with the movie site Netflix, the TV-based Hulu Plus, and other major software available. Music and other multimedia are heavily supported, too.

How to Read This Book

As you may notice, this guide is split into three parts:

- NOOK Tablet
- NOOK Color
- NOOK Simple Touch

If you have NOOK Tablet, just keep reading after this section. Have NOOK Color or NOOK Simple Touch? You'll want to hop ahead to the second or third sections. Don't worry. We'll cover everything you need to know about your device, so no need to read every part to get the information. And, at the very end of the book, you'll find a few helpful sections:

- Online resources
- Glossary
- Index

This official guide to NOOK is your best source, but there are also some great Web sites that can help you along, too. We list the best online resources, including Barnes & Noble's well-maintained site, Nook.com.

Don't know your Wi-Fi® from your GB? The glossary will set you straight with clear, thorough definitions of terms. Once you learn these words, getting the best out of your NOOK will be much easier. Finally, flip through the index to find topics quickly. If the table of contents provides the big-picture view, the index gives you the detailed info. And remember, anytime you feel you need one-on-one personal support with anything to do with your NOOK, you can always drop into a Barnes & Noble store. Their team of NOOK experts is always there to help.

Enjoy the book and feel free to follow us on Twitter at @browndamon and @alexanderbevier. We'd love your feedback!

Best,
Damon Brown and Alexander Bevier

NOOK
Tablet

1. A Look at NOOK Tablet

In the Box

Inside the NOOK Tablet box you'll find everything you need to get started. The start-up process will take you only a few minutes, and, aside from optional accessories, you don't need to make any other purchases. We'll get into NOOK Tablet covers and other fun additions later in the book, but for now let's crack open the box and see what's packed inside.

First, take off the wrapper. Then, bend the bottom half of the box to open the package. Here's what you'll see:

- Your NOOK Tablet
- A USB-to-micro-USB cord
- A wall plug

Don't see the cord and the wall plug? NOOK Tablet is at the top of the box, while the cord and the wall plug are snug inside the bottom of the container. Remove your NOOK Tablet and its accessories from the box.

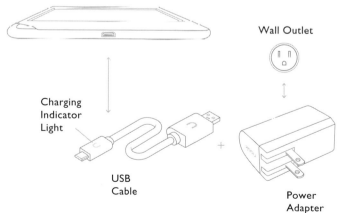

Wall Outlet

Charging
Indicator
Light

USB
Cable

Power
Adapter

NOOK Tablet accessories

NOOK Tablet

NOOK Tablet is less an eReader and more like a full-fledged tablet. It can play Flash video; run apps (including video and word games); and use interactive, pop-up style books. The memory can also be expanded so you can download as many apps, music, and reading material as you can handle. You'll notice that the dimensions of NOOK Tablet are hefty compared to other eReaders. It is actually intended to go beyond just reading books, so the plush color screen and solid weight make sense when you're watching videos.

The front of your NOOK Tablet displays the full-color touchscreen. Right below the touchscreen is the "n" symbol. One of the few buttons on NOOK Tablet is the "n" key found on the center-bottom of the device's front. Tap it when the power is on and your current options will pop up onscreen. Notice the curved hole in the lower left-hand corner? Please do not use it to hold your NOOK Tablet as it is a design element and is not reinforced.

Now tilt NOOK Tablet away from you so you can see the thin bottom right below the "n" key. Here is the hole for your USB wire. We'll plug it in after we finish our tour of the device. Turn the device to the right, past the hook, and check out the left-hand side. The lone button there is the power switch.

Turn the device to the right again and you'll be looking at the top of NOOK Tablet. Here you'll find a single hole for your headphones. NOOK Tablet has a speaker for playing music or other audio, but the headphone jack means you can enjoy your audio on your own as well. The headphone jack is 3.5 mm—the industry standard—so you can plug virtually any mainstream headphone into the device.

Turn the device to the right one last time and you will see the left-hand side. See the plus and minus buttons? These will control your volume. Once we turn on your NOOK Tablet, a little volume meter will appear onscreen. If you do decide to use headphones, you'll want to check the volume before you put them in your ears!

Now flip NOOK Tablet so the touchscreen is facing down. Check out the textured back. While other devices are smooth and, arguably, slippery, NOOK Tablet's raised back makes it easier to keep a grip while you read, watch, or listen. And, if you look very carefully, you'll see several rows of holes near the bottom edge. These holes are the speaker. It may seem small, but the speaker packs enough punch for your multimedia listening.

Finally, let's take a look at the microSD card. Keep your NOOK Tablet facedown and give the metal NOOK Tablet logo by the hook in the lower right-hand corner a pull. The hidden microSD card compartment will pop up.

You'll be downloading lots of books, periodicals and, perhaps, apps for your NOOK Tablet, all of which use a lot of memory. The microSD card allows you to expand the memory your NOOK Tablet can handle. We'll discuss microSD cards and download management further in Chapter 11, *Memory and Storage*.

Before You Start: Charge It

Okay, now that the tour is done, you'd probably like to start reading! It's hard to resist the urge to dive right in when you get a new device, but there is usually a step or two you have to take to guarantee a good experience. For NOOK Tablet Barnes & Noble recommends that you charge up your device before you start playing with it.

Let's get it charged before we take it for a spin. Take a look at the USB cord. One end has a traditional USB format that connects to most computers, while the other end has a micro-USB format that is usually used by devices. Plug the wider USB end into NOOK Tablet's wall plug.

The micro-USB end plugs into your NOOK Tablet. As you may remember, if you look at the "n" button, you'll find the micro-USB hole right below it. Connect the micro-USB in there and, finally, put your wall plug into an outlet. If everything is connected properly, you'll see a small orange light coming from the micro-USB end. The light will turn green once NOOK Tablet is fully charged.

Powering Up

Did the light on the micro-USB change from orange to green? Your NOOK Tablet is charged. We can finally turn it on! Hold the Power key, located on the left side of the screen. You should get a brief loading screen and then the intro page.

There are four steps to getting started here:

▶ Agree to Barnes & Noble Terms and Conditions.

▶ Set time zone.

▶ Connect to Wi-Fi.

▶ Register your device and default credit card.

You can also click on the intro video to get a brief personalized look at your device.

For now, let's go through the intro steps. The Barnes & Noble Terms and Conditions is the list of things you promise to do (and not do!) with the device. When you are done reading, tap on the Agree icon. After that, set your time zone.

NOOK Tablet Home Screen

Wi-Fi

The next big step is setting up your Wi-Fi connection. A Wi-Fi connection is required to finish setting up your NOOK Tablet. Are you in a Wi-Fi hotspot? A hotspot is any area where a Wi-Fi router is in range. Wi-Fi routers at public venues are usually free to join, but others, like the Wi-Fi router in your house, usually are (and should be) password-protected. It's unlawful to use someone's personal Wi-Fi without his or her permission.

The good news is that every Barnes & Noble store offers free Wi-Fi. It is a great alternative if you don't have your own Wi-Fi router at home.

If you are in a Barnes & Noble store, your NOOK Tablet will automatically detect the store's Wi-Fi network and ask if you want to connect to it. When you are in a hotspot, NOOK Tablet will ask you to choose a Wi-Fi connection. If you are at home, choose your personal Wi-Fi connection name from the list. If the Wi-Fi router requires a password, NOOK Tablet will ask you to type it in. A virtual keyboard will pop up onscreen that will let you punch in the letters and numbers in the password.

If you are having trouble with your home Wi-Fi showing up on the list of hotspots, try unplugging your router from the outlet and plugging it back in after five minutes. Check the list of NOOK Tablet hotspots again and it should be on the list.

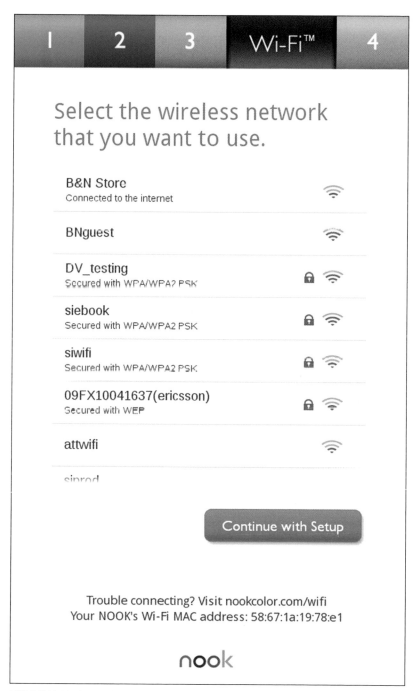

NOOK Tablet's Wi-Fi Settings screen

Account Information

The last step is registering your device, your account, and your credit card. Do you already have a barnesandnoble.com account? If you remember your username and password, type it in here. If you have an account, but don't remember the password, you can hop online and request a new password to be emailed to you.

If you don't have an account yet, go ahead and type in a username and a password that you will remember but others will have a hard time guessing. Barnes & Noble will check to see if anyone else is using the same username. As soon as you have an original username, you'll move on to the credit card.

Like other mobile devices, your NOOK Tablet allows you to make purchases without having to type in your credit card information each time. Instead, you type in a default credit card (it must be a credit card from a U.S.-based bank, with a U.S. address on the account) that will be stored privately by Barnes & Noble. Anytime you make a purchase, the amount will be charged to your card. Keep in mind that you can always change your default card or even input another card for a particular purchase—it's just much easier to have a credit card on file. Choose a valid credit card, type in the account number, and, once it is validated with a quick online check, you're ready to begin using your NOOK Tablet.

Navigating the Touchscreen

Your NOOK Tablet can understand six different types of gestures on its screen. If you move your finger up or down, NOOK Tablet will scroll through its menus. If you swipe your finger left or right, the pages will turn forward or back. If you pinch your fingers in or spread them out, the screen will zoom in or out.

You can also tap the screen to access a book, or to open up the book's menu. The menu will let you change the size of the font, find a specific word, or access the table of contents. You can also quickly tap the screen twice to receive more information about the book at the NOOK Store™.

The last feature on your NOOK Tablet touchscreen is the Press and Hold feature. Press and Holding a word while reading will let you highlight phrases, add annotations, or look up the word in the built-in dictionary. Also, Press and Holding on the front page or within apps will pull up a menu that changes based on what you're currently doing. We'll get into that further in the later chapters on book reading, periodical reading, and using apps.

The NOOK Button

The main menu is accessible by tapping the small "n" or NOOK button. These main menu icons are the most-convenient and primary way to navigate through the most used areas of NOOK Tablet. After tapping the "n" button, the following icons will appear

NOOK Tablet Quick Nav Bar

on the bottom of the screen.

> **Home:** This will take you to the Home Screen.

> **Library:** Quick access to the books, magazines, newspapers, comics, graphic novels, and apps on your NOOK Tablet and any

files you transferred to it from your personal computer.

▶ **Shop:** Takes you to the NOOK Store, where you can make online purchases.

▶ **Search:** Lets you look through both NOOK Tablet and the NOOK Store for anything you type in.

▶ **Apps:** Shows you all the apps installed on your NOOK Tablet.

▶ **Web:** Opens a Web browser for surfing.

▶ **Settings:** Accesses the Settings page to change several NOOK Tablet features.

The Status Bar

Like most devices, your NOOK Tablet includes an active area that gives you helpful info. There's no official name for it, so in this guide we refer to this area as the Status Bar.

Located at the top, the Status Bar shows you the following:

▶ The book, magazine, or item you were just using. Tap the name and it will open to exactly where you left off.

▶ Your history: Tap to get a listing of the last few items you were enjoying. You can select one of these and jump back in right where you were.

At the bottom of NOOK Tablet's screen, you'll see the following:

▶ The placeholder: A marker you left at a particular place in a particular book, magazine, or newspaper. Tap it and it will open up the reading material to where you marked it.

▶ Wi-Fi, power, battery status and clock

NOOK Tablet Status Bar

The Left Side of the Status Bar

The left side of the Status Bar is contextual: That is, it changes based on what you are doing. If you are reading a book, the status information will be much different from, say, if you were listening to music or surfing the Web.

There are several icons that can appear in the bar.

- Open Book
- Download arrow
- Email Envelope
- Pandora® "P"
- "n" update symbol
- NOOK Friends™
- Red Notifications circle
- Musical Note

The Open Book icon is a standard bookmark. If you open up your favorite book, magazine, or newspaper, and then go to another screen—such as the Home Screen—then your NOOK Tablet will keep your place. You can then tap the Open Book icon and your NOOK Tablet will take you right back to what you were reading. We'll talk more about reading in Chapter 3, *Books and Periodicals*.

The Download arrow means that your NOOK Tablet is currently getting a new book, magazine, newspaper, or app for you. With some devices, when you purchase an item you aren't able to do other things while you wait for the download. NOOK Tablet is great at multitasking, so you can start downloading something and, while it downloads, go back to your reading, Web browsing, or what have you. The Download arrow confirms that your download is still happening.

It will disappear when your download is complete. Don't worry if you're not sure how to download things yet, as we'll provide detailed instructions in the following chapters on *Books and Periodicals, Apps, Games,* and so on.

The Email Envelope means that you've got mail. NOOK Tablet connects to your email, so you can read, reply, and delete when you're on the go. The one thing NOOK Tablet doesn't do is actually "host" your mail—you'd need a regular email service, such as Gmail™, Hotmail®, Yahoo!®, or another service to run your email. NOOK Tablet just allows you to access it. That said, the Email Envelope will appear whenever you get new mail. You can sync up NOOK Tablet to check for new email, so you can stay connected, even while you're away from your home computer or laptop. More info on email is in Chapter 10, *Getting Social.*

The "P" stands for Pandora, the free music service that works well on NOOK Tablet. Available through the NOOK Store, Pandora streams music—that is, through the wireless Internet—into your NOOK Tablet. And, because NOOK Tablet is great at multitasking, Pandora can pipe music through the device while you're reading, Web browsing, or shopping for apps. You know it's playing when the "P" is in the corner. More info on music can be found in Chapter 6, *Music and Video.*

The "n" symbol popping up in the corner may look familiar, as it is the NOOK logo. It means that new software has already been downloaded to and installed on your NOOK Tablet. While the Download arrow means that NOOK Tablet is getting one of your requested books, periodicals, or apps, the "n" symbol means that your NOOK Tablet has downloaded new software from Barnes & Noble. If you're familiar with home computers, you know that hardware companies regularly update their software to eliminate bugs, make improvements, etc. NOOK Tablet uses the same system and, as on other devices, NOOK Tablet software updates are free.

The white icon with two people shows that activity is happening in NOOK Friends. Similar to Facebook®, NOOK Friends allows you to follow other people and see what they are currently reading, what apps they are using, and what their opinion is on the latest new reads. In turn, of course, you can share your current opinion and activities, too. A nice perk is that you can also find out what books your friends have available to borrow, plus you can lend books to your friends. When the NOOK Friends icon appears, there is new activity happening among your friends it could be a new person sending you a friend request, a message sent to you by a current friend, or another event pertaining to you and your friends. We'll discuss NOOK Friends more in Chapter 10, *Getting Social*.

The red Notifications circle means that you have something new that needs your attention. It could be a new message from a friend or a new book available through LendMe™. The number inside the circle tells you how many notifications you have. Tap the circle to see all the notes.

Finally, the red Musical Note icon means that your tunes are playing. In addition to displaying books, your NOOK Tablet can play your favorite music. You can transfer MP3s from your home computer, and listen to them through your NOOK Tablet. As we discussed above, your NOOK Tablet has a headphone jack at the top for standard earphones as well as a set of hidden speakers on its backside for public listening. When you are playing music, the red Musical Note will appear in the Status Bar. Like other activities, you can start up the music player and move on to reading, Web surfing, etc., while the music continues. We'll tell you how to get music on your NOOK Tablet in Chapter 6, *Music and Video*.

The Right Side of the Status Bar

The first icon, which looks like a fan, is your Wi-Fi icon. When Wi-Fi is active, you'll see little waves coming from the bottom to the top of the icon; if you have any other Wi-Fi-enabled devices, you are probably familiar with this symbol. If the fan is empty, however, then you don't have Wi-Fi right now. Remember that it doesn't mean that Wi-Fi isn't enabled, but that you aren't getting Wi-Fi right now—your local hotspot could be down, or there could be some other technical difficulty. If you're having trouble, try going through the Wi-Fi start-up process again, or skip ahead to Chapter 13, *Troubleshooting*, to get yourself squared away before proceeding.

Just above the Status Bar are quick links to Books, Newsstand, Movies, Music, and Apps.

The second icon, which looks like a little battery, is your Power icon. When NOOK Tablet is fully charged, the battery will look totally full. As the device is used, the battery will drain. No worries: Your NOOK Tablet can handle up to eight hours of continuous use and can sit on standby for days. However, when your power finally does get low, the battery meter will shrink and, eventually, flash to warn you that you need to recharge.

The third and final icon, clock, gives you the current time. Tap on it, though, and you can get a list of quick settings:

- Today's date
- Battery Life
- Wi-Fi switch

- ❯ Mute
- ❯ Auto-rotate screen
- ❯ Brightness

The battery life gives the actual percentage of remaining power. The Wi-Fi switch, labeled On and Off, allows you to turn off the wireless connection. Why would you want to? Having the Wi-Fi enabled actually takes more energy, so if you're low on battery power and don't need to download anything, you can turn it off. Also, if you're having challenges getting online, you can try turning the Wi-Fi off for a few minutes and then turning it back on. Again, check out Chapter 13, *Troubleshooting,* for more help with Wi-Fi challenges.

Mute turns off the sound completely on your NOOK Tablet. You can also tap the up and down volume controls to adjust it.

The Auto-rotate screen allows you to customize your view to suit the shape of what you read. Most books, magazines, and newspapers are meant to be read vertically, which is why your NOOK Tablet is taller than it is wide. However, many can be read horizontally, too. Reading with the vertical view means one page appearing at a time, but reading with the horizontal view means two pages appear across the screen (unless you modify it to read as one very broad column). The horizontal view is great for picture books and other publications that have multipage illustrations. We'll talk more about that in Chapter 3, *Books and Periodicals.* For now, know that NOOK Tablet will adjust to either a portrait or landscape setting depending on how you hold it. Toggle the Auto-rotate switch to keep the view fixed no matter how you hold the device. A lock icon will also appear momentarily when you open a book so you can decide on your orientation right away.

⚙ settings

Device Settings

Device Info 〉

Wireless 〉

Quick Settings ⚙

November 29, 2011

Battery 94%

Discharging

Wi-Fi ON

Connected to B&N Store

Mute

Mute sounds (except for media)

Auto-rotate screen ✓

Switch orientation automatically when rotating your NOOK®

Brightness ▼

Adjust the brightness of the screen

Shop 〉

Social 〉

Reader 〉

Search 〉

📷 📖 🛜 🔋 11:42

NOOK Tablet Quick Settings

NOOK Kids books are always shown in landscape format, no matter how you have your orientation set.

Finally, you can adjust screen brightness. Tap the Brightness icon and a sliding bar will appear onscreen. Use your finger to increase or decrease the light. The brighter the light, the easier it may be for you to see, but it also takes more battery power. Adjust it as you see fit, balancing your battery needs and your comfort level.

Advanced Settings

Notice the little cog in the upper right-hand corner of the Quick Settings menu? That contains the device and app settings, and a more complete list of controls. Here are the device settings:

- Device info
- Time
- Wireless
- Security
- Screen
- Keyboard
- Sounds

The app settings are right below the device settings:

- Home
- Shop
- Social
- Reader
- Search
- Power save

The Advanced Settings page is one of the most powerful pages; here you can tweak and adjust your NOOK Tablet to truly make it your own. The Advanced Settings page ends up touching on nearly every aspect of NOOK Tablet from your reading experience to your ability to connect with other NOOK readers, so we'll be referring to this screen throughout the book. At this point, just remember that you can access it either through the cog on the Quick Settings menu or on the Main Menu under Settings.

Warranty Options

Finally, before you start having fun with the device, you want to consider the warranty options. Barnes & Noble has a couple of different ones to choose from:

NOOK Tablet Protection Plan
(based on http://www.nook.com/warranty)

	Standard Warranty (1 year)	B&N Protection Plan (2 years)
Customer Service	x	x
Rapid Replacement	x	x
Accidental Damage		x
Extended Service		x

The best part is that you automatically got the basic protection as soon as you bought your NOOK Tablet! It's just a matter of deciding if you'd like to upgrade to the advanced protection plan.

Both plans include:

- Free customer support
- Rapid replacement if NOOK Tablet malfunctions
- Minimum of one year protection

However, the B&N Protection Plan adds:

- Two years of protection
- Rapid replacement for accidents like spills and cracks

Visit http://www.nook.com/tablet/warranty or call 1-800-843-2665 for the latest B&N Protection Plan pricing. For more details on the support plans, read the FAQ at the end of this book.

Now you have everything you need to get started with your NOOK Tablet. But what about reading books, downloading items, and managing your ever-growing collection? The rest of this section of the book is dedicated to making sure you get the most out of your device. Let's get started.

2. NOOK Store

While there are a few books—such as Bram Stoker's *Dracula*—included on your NOOK Tablet from the get-go, you probably want to read other books and expand your digital library. This is where the NOOK Store comes into play.

The NOOK Store is where you will be buying, sampling, and browsing through the vast array of books available for reading on NOOK Tablet. The shop features all of the books and categories you'd be able to find at a Barnes & Noble retail store. After all, what good is an eReader with nothing good to read? To access the NOOK Store, push the NOOK button and touch the Shop icon.

Be aware that while there are many free books in the NOOK Store (over a million free titles), many cost money. Buying them doesn't require you to re-input your credit card number, making new purchases quick and easy. You can set NOOK Tablet to require a password for every purchase by going to the Settings menu and tapping the Shop icon.

The NOOK Store Front Window

There are two main sections in the NOOK Store's front page: Book Genres and Popular Lists. If you're looking for a specific type of book, scroll through the Book Genres.

The Popular Lists pane covers similar ground. This section reveals greater detail about the current major titles in the literary world. Here, you'll find things like the *New York Times* bestsellers and new releases. It also lists featured books on sale.

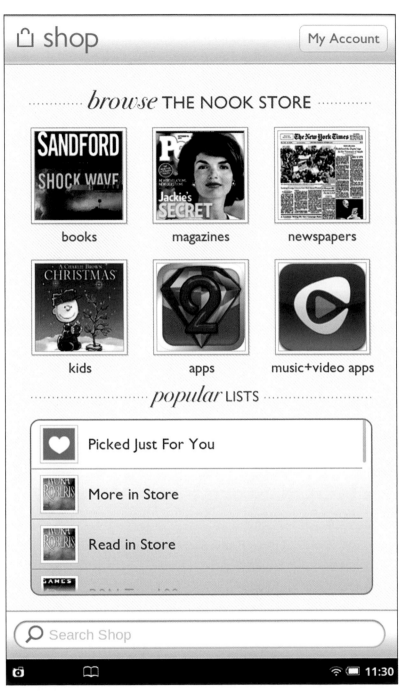

NOOK Store's front page

Browsing

Browsing through the Store is easy. By using lists, or simply by searching, countless books are waiting to be discovered. If you aren't sure what books to look for, it's best to start with a list, or simply search for a specific book you want. Just tap the list/section you want and start narrowing your search. If you don't have a specific title in mind, we'll tell you how to find a book through the first pane.

Books

If you already know what book you're looking for, then use the Search Bar at the bottom of the screen. Type in either the name of the book or the author, and you'll find a list of works containing the words you've typed. This function works like any other search engine. If you're browsing for something new and exciting to read, however, we recommend you go through the Store's detailed list to find something that piques your interest.

Tap the Books icon on the Store's Front Window and you'll see a page with a scrollable collection of lists at the top and several scrollable shelves of recommendations at the bottom. Each category covers a different literary genre. From biographies to humor, every style is represented.

Scan each page until you see a genre that looks interesting and select it. The next section will contain another series of pages detailing a more specific kind of book. For example, selecting Humor will lead to a list displaying the variety of humor books, like essays or cartoons.

Here, you'll find a countless supply of books directly related to your search. Browse until you find a book that looks interesting. Tap on a book's cover and a brief summary of the book and publication info will appear. If you want to buy it, tap the button with the price on it. Then, the button will change to say Confirm. Press Confirm, and the book will automatically be downloaded to your NOOK Tablet. If you aren't sure about buying the book, look to see if the publisher has made a free sample available, and if so, download it by pushing the button marked Sample. After you've decided whether or not to purchase the book or download the sample, simply press the Back button—the little button with an arrow pointing left—and look for other books.

> Later on, if you like the sample of what you've just read, you can purchase the complete text simply by tapping the Buy Now button at the top of every page.

Learning how to buy books is obviously central to enjoying your NOOK Tablet. After you've learned how to find books, try downloading a few samples. It won't take long before you have a full library of classics and new bestsellers.

In our next chapter, we'll talk about how to access and read all the books you've just sampled and purchased.

Periodicals

You can browse and shop for magazines and newspapers similarly to books. The difference, however, lies in purchasing them. Instead of being able to download a sample, with all publications you can try a free 14-day trial, or simply purchase the current issue. Once you are certain you want a particular magazine or newspaper, you can also subscribe to it for a subscription fee, and each new issue will automatically download to your NOOK Tablet as soon as it is released. There is usually a steep discount for subscribing compared to buying each individual issue, and then there's no stress about ever forgetting to download one. It guarantees a continuous supply of new content for your NOOK Tablet.

Apps

NOOK Tablet users can also download apps for their devices. These are programs that work like tools that tell the weather, or even games like Scrabble® or crossword puzzles. Navigate through the Apps Shop until you find one that sounds compelling, push the button with the price on it, then Confirm, and the app should be available to use within seconds. In order to know more about the variety of available apps, we've included an entire chapter about them later in this section.

3. Books and Periodicals

Now that you've purchased some exciting new titles and are on your way to building an impressive library on your NOOK Tablet, this chapter will show you how to dive into these great new reads and enjoy special reading features like bookmarking, looking up words, and more. We'll also discuss how to navigate and organize your Library by creating your own personal bookshelves and making books, magazines, and newspapers quick and easy to find.

Finding Books and Navigating the Library

The NOOK Tablet Library can be accessed by pressing the NOOK button to open the Quick Nav Bar and then tapping the Library icon. When you first tap the Library icon, a list will appear with all the reading you have downloaded to your NOOK Tablet. This list can be managed with the two navigation menus, found immediately under the Library title. The first menu helps navigate through books, periodicals, apps, and any files you have transferred from your personal computer; we'll explain this further in Chapter 11, *Memory and Storage*. The second menu reorganizes the list by title, author or the most recently added books in the list.

If you don't like the default layout of the Library, you can change it to look like a simple list of books. Near the top right corner of the NOOK Tablet Library screen, there are four buttons. Pushing any of them changes the appearance of the Library, so choose whichever one suits your preference.

To locate a book, simply scroll through the page until you find the book you want to read. To scroll, either tap the directional arrows at the bottom of the screen, or swipe your finger along the screen. Tap the book's cover and the book will open. You can also open magazines, newspapers, and PDF files by tapping on their covers.

If you're looking for a book you know you have, but it isn't in your Library, be sure to check out your Archives. Archived books are books that are not on your NOOK Tablet but Barnes & Noble recognizes that you own them. To get to the Archive, tap the My Stuff icon in the Media Bar and then tap Archived in the drop-down menu. Tap the button marked Unarchive to re-download items. To learn more about this, check out the Chapter 11, *Memory and Storage.*

Reading Books

Once you have opened a book, diving in to start reading is as easy as a tap or a swipe. Let's start with turning pages. The first way to turn a page is to tap along the right or left margin of a page, along the very right side or left side of the touchscreen. You can also turn a page by swiping your finger from right to left, or left to right, to turn the page back.

Once you master page turning, you'll be able to navigate through anything NOOK Tablet has to offer. There are, however, many other features inside a book that are worth discussing. NOOK Tablet lets readers annotate their books, and features several functions to expedite page-turning searches for specific parts of the book. You'll find more about other ways to personalize your NOOK Tablet in the next section.

Notes and Highlight

Sometimes, something you read is so striking, you'll want to remember exactly where you read it, and that might not seem like the easiest thing to do on an eReader. After all, you can't fold down a page corner on NOOK Tablet. Fortunately, NOOK Tablet has several key features that far surpass simple page-folding.

Notes and Highlight are very similar, except the former lets you add a few quick comments. Essentially, both features let you mark a word, or passage allowing you to access it from the Content page. This makes looking for specific parts of a book extraordinarily easy.

By holding your finger down for a second on the text in a NOOK Book, a horizontal menu pops up (the Text Selection toolbar) that lets you do several things to that selected text. You can highlight the text, add an annotation, share it with friends, post it on Facebook or Twitter®, or look a particular word up in NOOK Tablet's built-in dictionary.

Highlighting a word or passage helps mark key points of the book that you may want to return to later. Adding notes to the highlights also lets you add any comments you'd like to remember while reading the book. If there is a note on a page, it will be marked with a logo resembling a little note on the right side of the page. Tapping it will display the note you left.

Bookmarks

Bookmarks, as in printed books, mark an entire page for future reading. Bookmarks don't single out noteworthy words or allow annotations. Instead, they simply mark a page so you can easily return to it later.

When you tap a page in a NOOK Book, a small tab that looks like a ribbon with a lower case "n" appears in the top right corner. Tapping it highlights the Bookmark, and this page number will be saved for future reading. To undo a Bookmark, simply tap it a second time, and the ribbon should disappear.

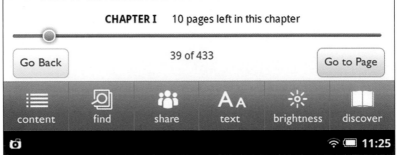

Dracula (Barnes & Noble Classics Series)

you are going to?' She was in such evident distress that I tried to comfort her, but without effect. Finally she went down on her knees and implored me not to go; at least to wait a day or two before starting. It was all very ridiculous, but I did not feel comfortable. However, there was business to be done, and I could allow nothing to interfere with it. I therefore tried to raise her up, and said, as gravely as I could, that I thanked her, but my duty was imperative, and that I must go. She then rose and dried her eyes, and taking a crucifix from her neck offered it to me. I did not know what to do, for, as an English Churchman,[i] I have been taught to regard such things as in some measure idolatrous, and yet it seemed so ungracious to refuse an old lady meaning so well and in such a state of mind. She saw, I suppose, the doubt in my face, for she put the rosary round my neck, and said, 'For your mother's sake,' and went out of the room. I am writing up this part of the diary whilst I am waiting for the coach, which is, of course, late; and the cru-cifix is still round my neck. Whether it is the old la-

CHAPTER I   10 pages left in this chapter

39 of 433

Go Back   Go to Page

content   find   share   text   brightness   discover

11:25

The Reading Tools menu

To access the page later on, go to the Content menu. The Bookmarks tab is the one closest to the right. Tap it, and a list of Bookmarks will appear on your screen.

Navigating NOOK Books

NOOK Books are extraordinarily easy to navigate. In addition to the above-mentioned Notes and Highlight, NOOK Tablet has several features to make finding the exact part of a book a simple as a tap.

In order to start, tap on the center of any page of the book, and you'll find the following reading tools:

- Slide bar
- Content
- Find
- Share
- Text
- Brightness
- Discover

Even with all of the in-book options, you can always tap the NOOK button and navigate NOOK Tablet like you normally would. This will take you out of the book and into other pages like Settings, the NOOK Store, or even back to the home screen.

Slide bar

Notice the horizontal bar at the bottom of the NOOK screen? This lets you navigate quickly to a specific page by simply moving your finger left or right along the bar; the book will either advance forward or back several pages, depending on how far you move your finger. Once you let go, NOOK will tell you some brief details about the page, and usually how many pages are left in that chapter.

Content

The Content button brings you back to the table of contents. Here, you can easily jump to each chapter in the book and navigate through the list in the first pane to see what each chapter is called and on what page the chapter begins. Tapping on the chapter name takes you to that specific chapter.

The Content screen also has tabs in the upper section of the page that you can use to navigate through your Content, Notes & Highlight, and Bookmarks. To look through them, find the other tabs near the top of your NOOK Tablet screen and tap them. The lists will look similar to the Chapter pane, but will feature all the Notes, Highlights, and Bookmarks you've already made in the book. These are great ways to personalize the navigation experience.

Find

The Find feature lets you look for any word in the book. Just type in the word and within seconds, you'll see a list of places where the word appears in the book. The list displays part of the sentence in which the word appears, as well as the page number.

To use the Find feature, open up the Reading Tools menu and tap the button marked Find. This will bring up a Search Bar. Type in the word you want to look for, tap the "Search" button in the lower right corner of the keyboard, and a menu similar to the ones found in the Content feature will appear, listing every appearance of the word you just typed. From here, tap the section of the book you want to go to and NOOK Tablet will take you there.

These features don't only work for NOOK Books. The Reading Tools menu works exactly the same way in PDF files (though PDFs saved as images aren't searchable) and most periodicals.

Text

There are five different options in the Text Settings menu. The first one—located in the middle of the screen—changes the font size. This changes the size of the words on the pages. The next tab changes the style of the font. This affects the visual appearance and flair of the typeface. Fonts include Trebuchet and Gill Sans. The following option allows you to change the color of the text itself. Next, the paragraph icons change the line spacing and the margins around each page. This changes how closely spaced the words are to each other and how close they are to the edge of the screen.

Adding books from places outside of the Barnes & Noble Store is a great way to add books to your NOOK Tablet. There are many free books available for your NOOK Tablet, including NOOK Books of works in the public domain (works that are no longer copyrighted) and other free eBooks available from places other than Barnes & Noble. (Note, however, that all eBooks must be in EPUB or PDF format to be read on a NOOK.) We'll talk about getting books from sources other than the NOOK Store later in this book, and address how to add them to NOOK Tablet in Chapter 11, *Memory and Storage*. We'll also talk about where to download free books in the *Online Resources* appendix.

The final option, the toggle switch on the bottom right, lets you set all of these options to the publisher's defaults (if provided). In most cases, the publisher's defaults set everything the way the book-publishing company wants it, setting the font, margins, and line spacing the way the publisher intended.

Creating Bookshelves

To best organize and manage your NOOK Library, you'll probably want to create your own bookshelves. This way, you can personally manage the books any way you want. To do so, start by going to the Media Bar in the Library, tap the My Stuff button and in the pull-down menu tap on My Shelves.

From here, tap the Create New Shelf button in the upper left. This opens a page that will let you name the new Shelf. We recommend that you use a name that will be easy for you to remember and understand later, such as a Shelf named for each literary genre you like or the month you purchased the book. Type in the Shelf's name, and tap Save in the bottom right corner.

After you name the new Shelf, it will appear at the bottom of the screen—if you don't see it right away, scroll the screen down with your finger. Tap the Edit icon and a list appears of all the content currently on your NOOK Tablet. Go through this list and check each book you want to place on that Shelf by tapping the check box next to the book's title. A check mark should appear in the box if you tapped it correctly. Once you've checked each book you want on the Shelf, tap Save, and the new Shelf will appear with your added books on it.

If you ever want to add anything to the Shelf, tap the Edit button right under the Add Shelf button. The Shelf screen only shows four you are going to?' She was in such evident distress that I tried to comfort her, but without effect. Finally she went down on her knees and implored me not to go; at least to wait a day or two before starting. It was all very ridiculous, but I did not feel comfortable. However, there was business to be done, I could allow nothing to interfere with it. I therefore tried to raise her up, and said, as gravely as I could, that I thanked her, but my duty was imperative, and that I must go. She then rose and dried her eyes, and taking a crucifix from her neck offered it to me. I did not know what to do, for, as an English Churchman,[i] I have been taught to regard such things as in some measure idolatrous, and yet it seemed so ungracious to refuse an old lady meaning so well and in such a state of mind. She saw, I suppose, the doubt in my face, for she put the rosary round my neck, and said, 'For your mother's sake,' and went out of the room. I am writing up this part of the diary whilst I am waiting for the coach, which is, of course, late; and the crucifix is still round my neck. Whether it is the old lady's fear, or the many ghostly traditions of this place, or the crucifix itself, I do not know, but I am not feeling nearly as easy in my mind as usual. If

39 of 433

The Text Selection toolbar

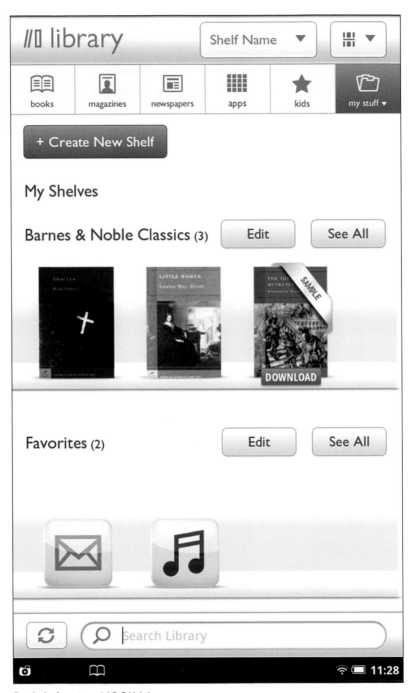

Bookshelves in a NOOK Library

books on its front page, but you can tap the See All button to see all the books on the shelf.

Borrowing Books

One of the biggest complaints about digital books is that, unlike traditional books, you can't borrow them. And one of the biggest benefits of using NOOK Tablet is that you can borrow books!

There are two ways to borrow books:

- From a NOOK Friend
- From your public library

With NOOK Friends, you can borrow books from your connected friends and lend books to them, too. All it requires is sending an invitation to a real-life friend via NOOK Tablet. To get connected, your friend needs to:

- Be a registered user at the Barnes & Noble Web site
- Have a registered NOOK device or NOOK eReader app including NOOK for iPad™, NOOK for Android™, NOOK for iPhone™, NOOK for PC™, and NOOK for Mac™.

To learn more about NOOK Friends, check out Chapter 10, *Getting Social.*

Periodicals

One of the best features of NOOK Tablet is its ability to download periodicals, giving you easy access to the latest newspaper and magazine issues. A new periodical you've subscribed to will download automatically when you're connected to Wi-Fi. The NOOK Store offers plenty of magazines and newspapers from cities big and small.

To purchase a periodical, navigate through the Store's menus— or simply search for a specific magazine or newspaper—and tap either Subscribe or Buy Current Issue. Subscriptions typically cost significantly less than the price of buying individual issues month after month. As discussed in the previous chapter, if you want to sample a magazine or newspaper before you subscribe, many periodicals offer a free 14-day trial subscription that will let you test-drive the publication before committing to a full-year subscription.

Reading Periodicals

By now, you're probably used to reading books on NOOK Tablet, and magazines aren't too different. After purchasing a magazine or newspaper, go to the Library and tap the one you want to read. The magazine's cover should be the first thing to appear.

Periodicals on NOOK Tablet are designed for ease of navigation. The biggest difference between periodicals and books is that newspapers have subsections, each of which has its own table of contents. After purchasing an issue of a newspaper, look through the table of contents for a section that seems interesting. Search through the pages of stories until you find one that you would like to read, tap the headline, and you'll be presented with the article. Once you've finished reading the article (it may be several pages long) the bottom of the last page will show Next Article and Previous Article buttons (which also appear on the first page of articles). Tap them to access the next or previous pieces. Or, you can return to the table of contents by tapping anywhere on the screen to access the Reading Tools menu, and then tap Content.

Archiving and Deleting Periodicals

Periodicals are updated regularly, so it's easy to be overwhelmed by the amount that can clutter up your NOOK Tablet's library. In order to avoid this, you can archive periodicals the same way you can archive a book. In your Library, if you double tap the periodical you want archived, you'll be brought to the Detail page, where after tapping "Manage," you will be given two options:

- Archive
- Delete

Archive will do just that—the magazine or newspaper will be removed from NOOK Tablet and placed in the B&N Archive until you want to re-download it, which we'll cover in "Managing Space," in Chapter 11, *Memory and Storage*.

Delete will remove the periodical completely from your device, storage, and your bn.com account. There will be no trace of it on your account aside from the receipt of purchase.

Unsubscribe from a Periodical

Unfortunately, there isn't yet a way to unsubscribe from a periodical directly from NOOK Tablet. The only way to do so is by going through Barnes & Noble's Web site. From the Web site, however, it is a very simple process to unsubscribe.

Simply go to BN.com and log in with your account, click on Manage Subscriptions, and click the Cancel Subscriptions button for each periodical you want to unsubscribe from. You'll be able to keep the issues that were downloaded during the time you subscribed, including those downloaded during the free trial.

4. NOOK Kids

One of the biggest benefits to having a NOOK reader is that the device brings rich, colorful books to life. Kids books are awesome on the NOOK Tablet, and Barnes & Noble designed the reader so it can take new and even classic titles to a whole new level.

The first step is downloading a NOOK Kids book. You can open up one of the NOOK Kids books preloaded onto your NOOK Tablet or, if you like, tap the "n" key, use the Quick Nav Bar to go to the Store, and search for a favorite title.

Once you get your NOOK Kids book, tap the cover to open it. Notice how it goes horizontal? Most NOOK Kids books (excepting some chapter books) are presented in landscape mode. Younger books usually have big, bright pictures, and the wider view allows better detail than the vertical setup. Turn your device so it sits in both of your hands.

NOOK Kids books can be read or read aloud

To the right is the full-color book cover. To the left there are three possible reading options:

- Read by Myself
- Read to Me or Read and Play
- Read and Record

Let's take a closer look at these three very different ways to read.

Read by Myself

Tap the Read by Myself option. You can now use your finger to slide from page to page: Push to the right to go forward and push left to go back to the previous pages. Again, all children's books show a two-page spread, so you can quickly go from one two-page section to the next. The two-page format makes perfect sense when you realize most NOOK Kids books have a page of text facing a page of illustrations or photos. Some books, like those from Dr. Seuss, have a combination of text and illustrations across both pages.

There is a little arrow at the bottom of the screen, nestled right where the two book pages meet. Tap this arrow and thumbnail images of all the book pages will appear in the bottom half of the screen. If you like, you can skim through the book quickly by sliding the thumbnails left or right with your finger. It is a great way to skip to your favorite part of the book.

Read to Me

Many NOOK Kids books also have the Read to Me option. Go back to the book cover, either by sliding right until you get to the front of the book or by tapping the white arrow and sliding the thumbnaiils to the cover. Once you get there, tap the Read to Me option.

Reading a NOOK Kids book

Now the book will be read to you by your NOOK Tablet! If you need to adjust the sound, turn the NOOK Tablet upright and use the volume up and down buttons on the right side of your device to make it higher or lower.

There is no standard narrator voice for NOOK Kids books, so each has its own unique sound. Try different books to see what the narrator will sound like.

The narrator will read what's on the two pages slowly and clearly. It will then wait until you flip the page. Want it to read the previous pages again? Just flip to those pages and the narrator will read them for you. You're in control of what you want it to read, so, unlike audiobooks, there's no reason to feel rushed.

Not every NOOK Kids book has the Read to Me option, but you'll be able to tell if it is available when you get to the cover page.

Read and Play

The final level of book interactivity is the Read and Play option. Again, if you're in the middle of a book, hop back to the cover page by sliding the pages or using the thumbnails. Tap on the Read and Play option.

At first Read and Play books seem like standard titles, but they have special spreads, marked by a star, that allow for cool interactivity. For instance, the book may ask you to tap the screen to touch a particular character or pinch an area of the screen to do a fun action. In these special sections, the pages will not progress with the traditional slide motions—the book will suspend page flipping so you can concentrate on the featured page spread.

Zooming in

Sometimes you want to get bigger text or get a closer look at a detailed picture. You can use your fingers to pinch the screen and zoom in or out. Touch the screen with your thumb and index finger, and spread them wide to zoom in. Do the reverse—push them closer—to zoom out.

Keep in mind that the zoom-in function doesn't work on the special Read and Play segments.

Read and Record

With NOOK Tablet, you can also record yourself reading a NOOK Kids book. Just tap the green Read and Record button on the opening page of the book, then the green Record button on whatever two-page spread you want to record, and start speaking into the Tablet's microphone. When you're done recording, just press the red Stop button or turn the page.

5. Web Browser

NOOK Tablet is great for reading books, magazines, and other periodicals. It can do much more than that, though. The rest of the NOOK Tablet chapters in this book will focus on its more advanced capabilities.

Surfing the Web

One of the most exciting features of NOOK Tablet is its full-fledged Web browser. NOOK Tablet can surf the Web as easily as your average computer.

Here are some details to keep in mind:

- You can only use the browser while connected over Wi-Fi.
- The keyboard pops up on the touchscreen.
- Unlike many other tablets, NOOK Tablet can play Flash video and animation.

To open up the browser, tap the NOOK button to open the Quick Nav Bar. The screen menu will pop up with the following options:

- Home
- Library
- Shop
- Search
- Apps
- Web
- Settings

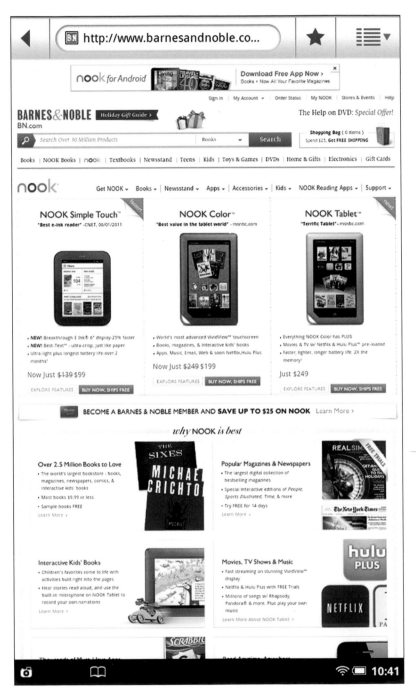

NOOK on the Web

Touch the Web icon and you're ready to surf. Feel free to hop online and play around with the Internet. Next, we'll take a closer look at the browser and show you how to get the most out of Web surfing on NOOK Tablet.

Browser Overview

If you're familiar with the World Wide Web, your NOOK Tablet Internet browser will be easy to use. In fact, it offers the same experience you'd have on your home computer or laptop. The only adjustments are the touchscreen keyboard and the screen (which can show the browser in both vertical and horizontal orientations). We'll discuss those differences in a second.

For now, let's take a tour of your Web browser. From the top, you'll see the following icons:

- Back Arrow
- Web Address Bar
- Bookmark Star
- List of Options

The Back arrow will open up the previous Web page. Don't worry about tapping it now, as you haven't been to any other Web pages, so there are no previous Web pages to open! If you tap it when there are no older Web pages to return to, NOOK Tablet will close the browser and send you back to your NOOK Tablet Home Screen.

The Web Address Bar shows the current Web page. You can type in your favorite Web address here. The Bookmark Star is a quick link to all your marked places. It also stores your most visited places as well as your history.

NOOK Tablet's Web browser

List of Options hides all the details you can tweak to personalize your Web-browsing experience. The full list of options will be discussed later in this chapter.

Your NOOK Tablet Web browser automatically opens up to the official Barnes & Noble NOOK page. Here you'll find lots of useful stuff, including interactive tutorials, descriptions of the latest NOOK Tablet accessories (which we'll touch on at the end of this section), and NOOK news via the NOOK blog.

There are three primary ways to browse a page:

- Tap
- Double tap
- Slide

Tap on an item, like a Web link or a video, and NOOK Tablet will open the Web page or play the video for you. Double tap your finger and NOOK Tablet will zoom in on a particular part of the page, or, if it is already zoomed in, will zoom out to show a bigger area.

Slide your finger up and down the touchscreen and the screen will scroll accordingly. Try touching the different links and exploring this Barnes & Noble page to get a feel for the browser. Next, we're going to visit your favorite Web pages. Like your computer's browser, there are two ways to get to a Web page:

- Do a Web search.
- Type in the Web address.

Tap the Web Address Bar and you'll see the keyboard pop up. Type in a topic or Web page name and your NOOK Tablet will use Google search to find it online. For instance, type in "Emeril Lagasse," press Go, and Google will give you a listing of everything related to the famous chef. It is literally the same information you'd get on a traditional computer, so you can feel comfortable that it is accurate and up to date. Also, if your topic is popular enough, Google will offer suggestions on what you are looking for as you're actually typing in the topic.

You can also type in a specific Web address. Touch the Web Address Bar at the very top of the screen and the NOOK Tablet will allow you to input the location. It is also smart enough to know that a "http:// www." belongs at the beginning of most Web addresses, so you won't need to type that part in most of the time. For example, type in "nytimes.com," and NOOK Tablet will display the home page of the popular *New York Times* Web site.

Browser Details

Now we'll look at two ways NOOK Tablet can make your Web experiences better.

Magnified Type

Whatever Web site you are on, slide your finger on the touchscreen. Notice the "+/-" symbols appearing in the lower right-hand corner? Try touching the "+" symbol. The size of the Web type will increase. Touch the "-" symbol and the Web type will get smaller.

Bookmarking

Bookmarking is equally important. Touch the Bookmark star located at the top of the screen, right next to your Web Address Bar. Your NOOK Tablet will show you three tabs:

- Bookmarks
- Most Visited
- History

The Bookmarks tab shows you all your favorite places on the Web. You don't have any bookmarks yet, but you can add one right now. NOOK Tablet will suggest different bookmarks, including your current Web page. If you want to add the Web page, look for the icon showing the name of the page and tap the Add icon. NOOK Tablet will ask if you have a special name for the Web page and to confirm the Web address. Touch OK and the Web page will be bookmarked for you. Tap the View option in the lower right-hand corner to toggle between the word list view and the visual thumbnail view of your bookmarks.

Most Visited shows the places you've browsed most frequently. Tap on one of them to visit the page now. See a page you'd like to add to your bookmarks? Just tap the Bookmark Star next to the Web address and your NOOK Tablet will add the page to your bookmark list. You can also tap History in the lower right-hand corner to clear your recorded Web browsing history.

The History tab gives a complete listing of every Web site you've visited. Again, tap the Bookmark Star next to any Web site and NOOK Tablet will bookmark it for you. You'll notice that the list has headers like Today. NOOK Tablet organizes your history based on day. Tap the Today headline and NOOK Tablet will hide the history behind the headline so you can easily see other days. As on the Most Visited page, you can tap the History option in the lower right-hand corner and clean out your Web browsing history.

Advanced Options

There are also a lot of advanced options, all hidden underneath the List icon in the upper right-hand corner. Get back to your Web browser, tap the List icon, and you'll see these options:

- New Window
- Bookmarks
- Windows
- Refresh
- Forward
- More Options

Like browsers on other computers, your NOOK Tablet Web browser allows you to have multiple pages open. Tap the New Window option, and it will open a new browser page.

Don't worry: The previous page didn't disappear! NOOK Tablet just tucked it away so you can explore a new page. Tap the Windows option and you'll get a list of all the windows you have open. Hit the "X" icon next to any of the open windows and NOOK Tablet will permanently close it. Tap on the name of the Web page to reopen the window.

Refresh will go online and make sure your browser is updated with the latest info from the Web site you're viewing. For instance, if there is breaking news on the *New York Times* Web site, you can tap Refresh to have your NOOK Tablet update the page with the most current reports.

The More Options section reveals additional selections:
- Add Bookmark
- Find on Page
- Page Info
- Downloads
- Settings

Find on Page allows you to look for a specific term. Tap it and the keyboard will pop up. Type in a term that you think appears on your current Web page, and NOOK Tablet will find it. Next to the term, your NOOK Tablet will show the number of matches as well as forward and backward arrows so you can jump between the different places where the term appears within the text. If the term doesn't appear in the text, NOOK Tablet will say "0 matches."

Page Info gives you the name, Web address, and any other public information about the current Web page.

Downloads lists any media you have downloaded onto your NOOK Tablet. It will give you the status on all of them, including how much longer it will take to download the item.

Finally, Settings sends you to the browser portion of the Settings menu located on your NOOK Tablet Home Screen. The options here include Text Size, Image Details, and Zoom.

6. Music and Video

The previous chapter discussed Web browsing, one of the many NOOK Tablet benefits that goes well beyond reading books and magazines. Another huge perk is that your NOOK Tablet can play some of your favorite multimedia. Multimedia means most of your favorite music and videos.

What you'll need is:

- Your NOOK Tablet power cable
- A PC or Mac with a USB port

Transferring Music and Video

Ready to get some songs and videos onto your NOOK Tablet? Before you do, keep in mind that your NOOK Tablet can't understand every file you throw on there.

Files That Work

Your NOOK Tablet can play these types of music files:

- AAC
- amr
- mid
- MIDI
- MP3
- M4a
- oog
- wav

NOOK Tablet cannot read WMA (Windows Media) files.

As far as video, your NOOK Tablet understands these formats:

- Adobe Flash
- 3gp
- 3g2
- mkv
- mp4
- m4v
- MPEG-4
- H.263
- H.264

It cannot read:

- MOV/qt
- AVI
- Xvid/DIVX
- WMV/VC-1

You can always use trial-and-error to see if your NOOK Tablet will read a file, but there's a quicker way. Turn on your computer—PC or Mac, it doesn't matter—and find the file you'd like to enjoy on your NOOK Tablet. If you have a two-button mouse, highlight the file and click the right mouse button. You will see an option called Get Info on a Mac or Properties on a PC. Select that option and you'll get a list of details about the file. See File Type and you can check if your file is on any of the lists above. If you really need to make a file playable on your NOOK Tablet, we recommend going online and downloading a file converter so you can turn the current file into one that's compatible with your device. There are dozens of free converters available on the Web.

Getting Your Files on NOOK Tablet

Once you find a file you can transfer, here's the process:

- ⫸ Disconnect the power plug from NOOK Tablet's power cable.
- ⫸ Plug the power cable into your NOOK Tablet.
- ⫸ Plug the other cable end into your computer's USB port.
- ⫸ Wait for the disk icon MyNOOK to appear.
- ⫸ Drag and drop the file(s) into the appropriate subfolder in My Files.
- ⫸ "Trash" the NOOK icon and disconnect NOOK Tablet from your computer.

Let's go through each step. First, remember that NOOK Tablet's power cable comes in two pieces: the plug and the cable. You'll now want to disconnect them. Grip the power cable right where it meets the power plug and give it a good tug. Put the power plug to the side for now.

Second, plug the opposite end of the cable into your NOOK Tablet. Nothing fancy here, as this is the same thing you do when you plug in the cable to charge up your device.

Third, connect the remaining end (where the power plug was just connected) into your computer's USB port. As you can tell, the USB plug has a thin, wide base. Look for the corresponding hole on your computer. If you have an older PC with a tall tower, you probably have at least one USB port in the front and in the back. On most modern desktop PCs and Macs, the USB ports will be in the back of the computer. And if you're a laptop user, the USB ports will be on either the right or left side of the computer.

Fourth, you'll know immediately if you've found the right hole and plugged it in securely: Your NOOK Tablet will light up and a drive

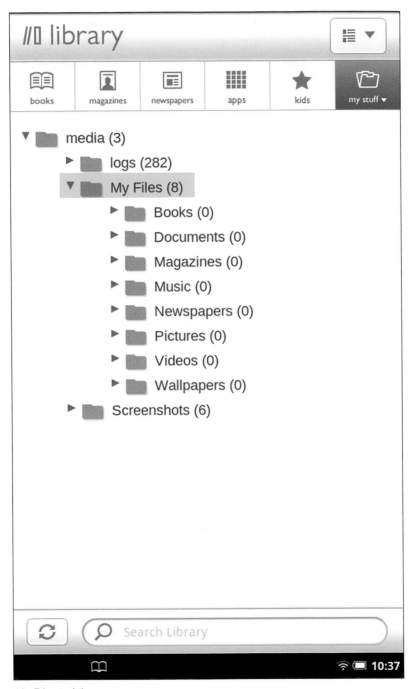

My Files in Library

icon called MyNOOK will appear on your desktop. Use your mouse to double click on this NOOK symbol. It will show several folders, but you are only concerned with the My Files folder. Here is where you can drag and drop your music and video files into NOOK Tablet.

Go deeper into the file and you'll see the following folders:

- Books
- Documents
- Magazines
- Music
- Newspapers
- Pictures
- Videos
- Wallpapers

Fifth, drag and drop your music or video file into the appropriate folder. Remember to ensure that the file type is included on the list of compatible types outlined earlier in this chapter.

Sixth, eject NOOK Tablet from the computer. It's not a matter of just unplugging it, as the computer needs to be given a heads-up that you're removing the device from it. It depends on your computer, but most PCs and Macs allow you to drag and drop the actual NOOK icon from the desktop to the Trash Can/Recycle Bin icon. This lets the computer know that you are going to unplug the device.

Now your new files are in your Library. To get there, tap your NOOK button, then touch the Library icon. You'll see a listing of items:

- Books
- Magazines
- Newspapers

- Apps
- Kids
- My Stuff

Tap on My Stuff. It organizes your files accordingly:

- My Shelves
- My Files
- LendMe
- Archived

Tap the square icon in the upper right-hand corner to toggle between the visual thumbnail view and the word-focused list view.

Listening to Music

You can listen to two types of music on your NOOK Tablet:

- Your music
- Streaming music

Your music consists of the tunes that we've just dragged and dropped into NOOK Tablet from your computer. However, since your NOOK Tablet is connected to the Internet, it can also stream music. Streaming music means your NOOK Tablet will download the songs live while you listen, kind of like an Internet version of the radio. Pandora and other music-streaming apps make it easy (and usually free!) to stream music.

Before we get into Pandora's streaming music, take a look at how you can listen to your own music catalog.

You can also access your music through the Music Player app.

Listening to Your Music

Now that you're in the My Files folder, tap on the Music folder. You'll see your music listed.

Tap on a song. If it is the right file format, it should start playing immediately. The song will play through the built-in speaker, but make sure to adjust the volume before putting on any headphones— you don't want to hurt your ears! The volume can be adjusted with the two Up and Down buttons on the right side of your NOOK Tablet, just below the headphone jack. If the volume is turned down, you can tell music is playing when a note icon appears in the lower left-hand corner of your screen.

The majority of the screen is taken up by the song or album cover art. You'll notice several icons immediately below the cover:

▶ Shuffle

▶ Repeat

▶ Album Cover View

▶ List View

▶ Browse

Indicated by the weaving arrows, the Shuffle icon toggles on and off. When Shuffle is on, your NOOK Tablet will mix up the order of the current list of songs.

The circular Repeat icon has three modes: Off, On, and Single. If Repeat is off, the music will stop once the song or album ends. If Repeat is on, the current song or playlist will start over again. If Single Repeat is on, it will repeat only the current song.

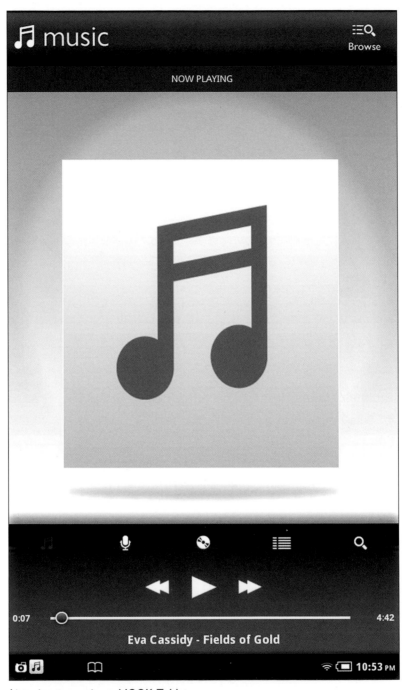

Listening to music on NOOK Tablet

Album Cover View and List View will toggle between the album cover art and the text-based list of songs. Tap the magnifying glass and you can browse for a particular song or album in your Library.

Below the first row of icons is a second set of controls:

▶ Rewind/Restart

▶ Play/Pause

▶ Fast Forward/Skip

▶ Timeline

Like a CD or DVD player, NOOK Tablet allows you to move forward or back within the current song by using the Rewind, Play, and Fast Forward buttons. Tap the Play button to pause or resume the music. Holding the Rewind or Fast Forward button will skip a few seconds back or ahead in time, while a quick tap of either will restart or skip the current song.

If you want more control, use the Timeline located below these icons. Touch the moving dot. Now you can slide the dot up and down the Timeline to go to a particular part of the current track. The time remaining and the total track time appear next to the Timeline. You'll see the name of the current track below the Timeline.

You can also make playlists using the music on your NOOK Tablet. Switch over to your List View (the fourth icon right below the album art). Now, find a song you would like in your new playlist and hold your finger down on the name. The following menu will appear:

▶ Play

▶ Add to Playlist

▶ Delete

▶ Search

Select Add to Playlist, then select New. Your NOOK Tablet will ask you to name the playlist. For instance, you might name a playlist Stephen King Reading Mix. Now you can find another song you'd like to add to your playlist, hold down your finger on the name, and add it to the Stephen King Reading Mix.

To access your playlist, tap the Browse icon in the upper right-hand corner, then touch the List View icon (the fourth icon below the main music screen). Find your playlist and tap on the name. Now your current playlist will appear. Tap the first song title to start the playlist.

Not happy with the order of the songs in a playlist? Find the song you'd like to move, hold your finger on the triple-line icon to the left of the song title, and drag it to where you'd like it to be in the playlist.

Streaming Music with Pandora

As we mentioned earlier, your NOOK Tablet can access music through the Internet, too. Here's what's required:

- ❥ A reliable Internet connection
- ❥ A music-streaming app, such as Pandora

If you're not sure how to download an app to your NOOK Tablet, flip to Chapter 7, *Apps.*

Why do you need a reliable Internet connection? Because the music isn't actually on your device. When you were listening to your own music, you literally transferred the music from your home computer, via a USB cable to your NOOK Tablet. With streaming music, your NOOK Tablet doesn't keep the songs on the device. Instead, like a radio, the device tunes into a station and just broadcasts whatever is playing. Unfortunately, that means that you can't listen to streaming music at, say, the beach unless you have a nearby Wi-Fi connection.

The second thing you'll need is a music-streaming app. NOOK Tablet doesn't have a built-in radio, so you can't just tune in to a particular signal like you would in your car. Luckily, NOOK Tablet can access a wide array of apps that will find stations for you. Most of them are free, too. Here are two of the most popular ones:

▶ Napster

▶ Pandora

Napster is a revamp of the well-known music software from the late 1990s. Now a legal service, Napster has more than 12 million songs available for little to no money.

Pandora is a more recent app. The service asks you for information on your favorite music. It will then play a song for you, which you can say you like or dislike. The service keeps playing songs until it is able to detect a pattern and—voila!—it knows what music you like with shocking accuracy.

Barnes & Noble is a big fan of Pandora to the point where it is included on your NOOK Tablet. No need to visit the online Apps Shop to find it.

To get to Pandora, tap the NOOK button on the front of your NOOK, and tap the Apps icon in the Quick Nav Bar. You'll see the blue-and-silver Pandora icon in your app collection. Touch the icon to get started.

Pandora is initially a free service, but you do need to create an account. If you are already a Pandora user through the Web or another device, go ahead and type in your registered email address and password. If you aren't a user yet, tap the Create New Account option and it will ask you for the following info:

- Preferred email address
- New password
- Birth year
- Zip code
- Gender
- Join newsletter
- Agree to Terms and Conditions

The newsletter isn't necessary, but it can be helpful to get tips and tricks when you first get started. However, you do have to agree to Pandora's Terms and Conditions, which can be read by clicking on the live link under Terms and Conditions.

Once you put in the information, Pandora will ask you for a favorite song, artist, or composer. For instance, if you really like jazz trumpeter Miles Davis' classic album *Kind of Blue*, you can type in "Miles Davis" (artist) or "All Blues" (a song from *Kind of Blue*). Here's where it gets interesting: Pandora usually won't play anything from *Kind of Blue* or even necessarily anything from Miles Davis, but it will play music similar to or inspired by the song or artist. For instance, Pandora may play a song from Miles Davis' friend and fellow band member John Coltrane.

Now that a song is playing, you can help Pandora make your personal station better. The screen will show you the following info:

▶ Album cover

▶ Song length and information

▶ Name of song

▶ Name of artist

▶ Name of album

▶ Control icons

▶ Station icons

The song length may be self-explanatory, but try tapping the small "i" icon next to it. The album cover will flip and Pandora will tell you exactly why it chose to play the song you're listening to. Depending on your music knowledge, you may not know what "syncopated drums" or "challenging guitar riffs" mean, but Pandora recognizes that pattern in the songs you like.

The Control icons lie right beneath your album info:

▶ Thumbs Up

▶ Thumbs Down

▶ Bookmark

▶ Play

▶ Skip

Thumbs Up means you like this song, while Thumbs Down, of course, means you don't like it so much. These two simple icons are the key to Pandora. Thumbs Up a song and Pandora will analyze the details about the song: the rhythm, the length, the vocals, and so on. It will then compare it to another song that you liked and see if there are any similarities. Now, the next song Pandora plays will, at minimum, have at least one quality shared among the previous songs that you liked.

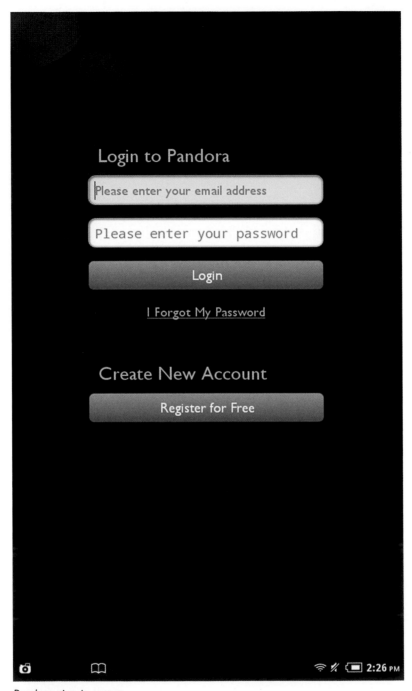

Pandora sign-in screen

Thumbs Down is essentially the reverse. If you don't like a song, Pandora will match it up to other tunes you gave a Thumbs Down to and see the similarities. It can be a lot trickier to avoid certain traits completely—like having a rock track with no guitar—but Pandora will try to minimize the incidence of traits you seem to dislike in its music selection.

Bookmark means you'd like to remember the song playing. Because of licensing restrictions, you can't rewind a song in Pandora or even play it again. In other words, Pandora isn't the same as your mobile music player or even your NOOK Tablet. What you can do is Bookmark a song. Once it is bookmarked, Pandora will give you a link so you can buy it from a digital music distributor.

Play is just like your NOOK Tablet music player. Tap it to play the music or pause the music. You'll want to utilize this icon since you can't rewind or replay songs in Pandora.

The Skip icon will let you jump to the next song. Pandora will only let you skip a certain amount of songs per 24-hour day, so be judicious in choosing what song you absolutely want to skip. Worst-case scenario, you can turn the music down or remove your earphones until the song is done.

The station icons are just below your Control icons:
- My Stations
- Add Station
- Sign Out

Tap the My Stations icon and you'll get a list of your stations. Pandora allows you to create multiple stations, so you could have one dedicated to, say, sports anthems and another one focused on opera. Once you have a collection of stations, use your finger to scroll up and down the list. Tap a station to start it up. The Edit icon at the bottom allows you to delete a station: Touch Edit, then tap the "-" symbol next to any station to remove it.

Be careful with stations, as once you remove one, it's gone forever—and some stations have dozens, if not hundreds, of your ratings saved within them!

Add Station will allow you to create a new station based on a particular artist, song, or composer. It will be added to your My Stations list.

Lastly, Sign Out will log you out of Pandora. If multiple people use your NOOK Tablet, you can log out and allow others to use their own Pandora accounts on the device.

Watching Video

Once you master playing music, watching videos is even easier. Again, go back to the My Files folder. Tap on the Video folder and you'll see your videos listed.

Tap on your video and, if the format is compatible, it will start playing. You can use the built-in speaker or headphones, both of which can have their volume adjusted by using the two Up and Down buttons on the right side of your NOOK Tablet.

With your video on the screen, you will immediately see four controls:

- Back (located in the upper left-hand corner)
- Rewind/Restart
- Play/Pause
- Fast Forward/Skip

Tap the Back button and NOOK Tablet will take you out of the video and back into the My Files section.

Rewind and Fast Forward will do just that when you hold them, but they will also restart or end the video if you tap the respective button. Finally, tap the Play button to pause, and tap it again to continue the video.

Video on NOOK Tablet

7. Apps

It's hard to hear about a device without someone talking about its apps. Short for applications, apps are simply the software programs used by a device. They are downloaded directly onto your device from an online provider; if you go to a physical store and buy it in a box, it isn't an app. Aside from that, apps are just software that allow your NOOK Tablet to do cool things.

NOOK Tablet has thousands of apps available right now. Most exceptionally, there are a large number of free apps, making it easier to take an app for a spin before you decide to keep it.

The categories are:
- Lifestyle & Interests
- Children
- News & Weather
- Education & Reference
- Productivity
- Entertainment
- Social
- Games
- Themes
- Health and Fitness
- Tools & Utilities

As you may imagine, Games and Productivity are two of the most popular app categories. We'll spend some significant time discussing those in chapters 8 and 9.

The Apps Screen

To get to the Apps menu, tap the NOOK button on your device and open up the Quick Nav Bar. As you may remember, your menu will have the following options:

- Home
- Library
- Shop
- Search
- Apps
- Web
- Settings

You, of course, want to tap Apps. The Apps screen will now appear.

The Apps screen has a clean, but effective, design. We'll start from the middle. In the main section you'll see the apps you've downloaded. Already see a bunch of apps there? Barnes & Noble pre-loads your NOOK Tablet with some apps it thinks you'll love. In fact, while a couple are just fun and games, some of them are absolutely essential for you to get the very most out of your tablet. In this chapter we discuss the pre-installed apps in depth, but know that the center of this screen will change based on your tastes and needs.

Just above your collection of apps is a bar reading, "More Apps You'll Love. SHOP NOW." The Discover Bar is the gateway to the apps in the NOOK Store. We'll talk about all the cool shopping later in the chapter.

For now, look at the lower left-hand corner for the Sync icon (looks like two arrows making a circle). Apps are software, and the creators

NOOK Tablet's main Apps screen

are always making updates to fix bugs, improve user experience, and even to add content. The frequency of these updates really depends on the app creator, but it's important to check regularly to make sure you have the latest version of an app. When you tap the Sync button, your NOOK Tablet will hop online and update the apps with their latest versions. You need to be online to sync your apps, so be sure to take advantage of it while you are in a Wi-Fi hotspot.

The Apps on Your NOOK Tablet

NOOK Tablet comes with a handful of cool apps pre-installed, so you can begin playing with apps immediately. As of winter 2012, here's what you'll get for free on your app screen:

- Chess
- Pandora
- Sudoku
- Netflix
- Hulu Plus
- Contacts
- Crossword
- Email
- My Media
- NOOK Friends
- Music Player

Chess and Crossword are digital versions of the classic board game and Sunday newspaper diversion, respectively. Sudoku is the popular number game. These are just a few of the games available for NOOK Tablet, which we'll get deeper into in Chapter 8.

My Media manages your pictures. Believe it or not, NOOK Tablet can also be a handy picture viewer, capable of full-screen displays, slideshows, and light photo editing. If you tap on the My Media icon, NOOK Tablet will show you all the pictures you have on file. NOOK Tablet comes with more than a dozen interesting photos pre-installed. Use your finger to scroll and you can coast up and down My Media. See a picture you like? The photos you're viewing now are called thumbnails, or miniature versions of the full-sized photos shrunk for the sake of navigating. Tap the photo, however, and you'll get the full picture in all its detail. We'll save the more detailed photo options, including editing, for Chapter 12, *Advanced Techniques*.

The Music Player app allows you to play songs transferred from your computer. We talked about it in Chapter 6, *Music and Video*, along with Pandora, the popular Internet radio service.

Contacts and Email apps turn your NOOK Tablet into a mini-computer. In fact, you can take care of quick work tasks just by using these two apps.

Finally, NOOK Friends enables you to connect with others who are also using NOOKs or NOOK eReader apps. You can look at it as a little social network for NOOK Tablet users. We'll go into detail about NOOK Friends in Chapter 10, *Getting Social*.

The Apps Shop

The handful of apps on your NOOK Tablet is a good start, but there are thousands of other apps that are worth checking out.

Furthermore, there are dozens more coming out every week. The place to get them is the online store, which we touched on in Chapter 2, *NOOK Store*. Now we'll go further into the apps section.

Tap the "n" button, get the Quick Nav Bar, and tap Shop. Here you'll find the books, periodicals, and other items. Tap the apps icon.Similar to the NOOK Store page, the Apps Shop shows the following parts:

- App Categories
- Top Picks in Apps
- Top Picks in Game Apps
- What's New in NOOK Apps
- Browse and Search Bar

Starting from the top, App Catagories lists all the different types of software available. You can use your finger to scroll the list of catagories up and down. If you're looking for a specific type of app, you can tap one of these categories and NOOK Tablet will narrow down your search.

Below the app list are Top Picks in Apps, Tops Picks in Game Apps, and What's New in NOOK Apps. There are icons for the apps, but, unlike the vertical categories list, the three app lists go horizontal. Use your finger on any of the lists and push it to the left or right to move it. You can also tap the See All option on any of the lists to see all of them listed vertically.

Go ahead and tap one of the app categories or the See All option on one of the lists. You'll see all the apps in that category. In the upper right-hand corner, you can choose your app layout from Picture, Picture and Text, and Text-free. The Picture option makes the app icons big and limits the details to the title, company, rating, and price. The Picture and Text option has those same details, but also gives a brief synopsis of the app. Finally, the stripped-down Text-free option removes the synopsis and shrinks down all the app icons, too. The Picture option emphasizes the visual, the Picture and Text option gives a good synopsis, and the Text-free option puts as many app choices as possible on one screen. Play around with them and decide which one works best for you.

Once you choose the type of layout you want, take a good look at the apps selection. Depending on the layout you choose, the apps will have the following details:

- Picture
- Title
- Company
- Rating
- Price and Purchase button

The picture is the logo or image for the app. You'll see this picture on your apps screen after you download the software, and it will be what you click on to start the app. The title and company are the name and the creator of the app.

The rating is how users have ranked the app. NOOK Tablet uses a five-star system. You'll notice a number in parentheses right after the star rating. It represents the number of people who have rated the app. Why is it important? The more people who have ranked the app, the more likely that the ranking will be representative of the app's quality. For example, if only one person ranked an app, that person may have a certain extreme bias. By telling you how many people have rated it, NOOK Tablet makes it easy to determine how much weight you should put on the current ranking.

The Price and Purchase button gives you the current cost and the option to download the app. If you decide to buy the app, the money is automatically deducted from the credit card you submitted when you first registered your NOOK Tablet. We'll have some fun downloading apps in the next section.

Finally, check out the Search Bar at the bottom of your touchscreen. Tap the icon and, near the bottom, tap Apps.

Let's take a look at the Search Bar. Tap the Search section and it will open up the Search page.

Go ahead and type in the name of an app with the keyboard. Let's try Aquarium Live Wallpaper, a bestselling app that turns your NOOK Tablet screensaver into a lifelike fish tank. You'll notice that, as you type, the Suggestion Box will fill up with several different apps that fit the name. You can now tap one of the suggestions and open up that particular app in the NOOK Store.

Trying to remember something you searched for a while ago? Once you do a search, NOOK Tablet will remember the search terms you used. These are shown before the search suggestions so you can pick up where you left off during a previous search.

Downloading Apps

Now that you know how to browse, why don't you find an app you like? When you find an interesting app, tap on the icon. You'll get a detailed overview of the app with this info:

- Name
- Company
- Version
- Rating
- Price and Purchase icon
- Add to Wishlist
- Share
- Overview

- Customer Reviews
- Screen Shots
- More Like This

The Name, Company, Rating, and Price and Purchase icons give the details we talked about earlier in the chapter, while Version tells you which version number is available for download. As app creators update the software, they will increase the version number. Version 1.0 means it is the very first version released to the NOOK Store, and incremental numbers mean revisions. If you sync regularly, you'll always have the latest version of an app. Some apps have two additional options: Add to Wishlist and Share. Add to Wishlist will put the app on the record as something you want. Your Wishlist can be kept for your own memory or shared with friends, as we'll discuss in Chapter 10, *Getting Social*. Share allows you to tell your friends about the app via your NOOK contacts, through Facebook, or on Twitter.

The remaining details of Overview, Customer Reviews, and Screen Shots, are at the top of the screen, and and More Like This is available at the bottom. Tap the tab to see the info you'd like.

Overview gives you a longer description of the app. It usually is a long paragraph written by the app creator. It can also include more info on the creator, including its Web address and contact info.

Customer Reviews lists all the reviews of the app with the review title, the user name, the rating, the posting date, and, of course, the review itself. Use your finger to scroll up and down the list of reviews. NOOK Tablet lists the ten most recent reviews, but additional reviews can be read by touching the More icon at the bottom of the listed reviews.

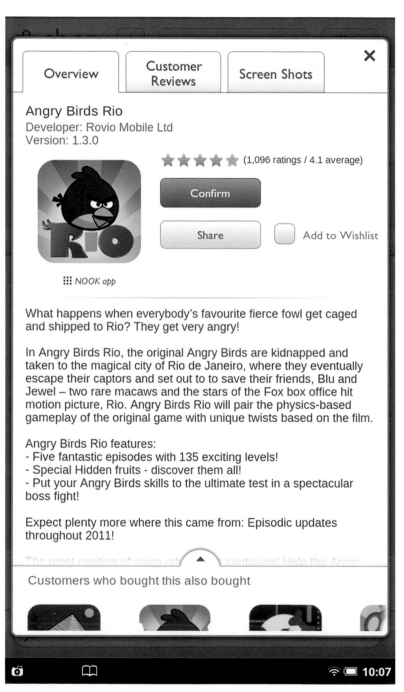

Angry Birds Rio
Developer: Rovio Mobile Ltd
Version: 1.3.0

(1,096 ratings / 4.1 average)

Confirm

Share   Add to Wishlist

⠿ *NOOK app*

What happens when everybody's favourite fierce fowl get caged and shipped to Rio? They get very angry!

In Angry Birds Rio, the original Angry Birds are kidnapped and taken to the magical city of Rio de Janeiro, where they eventually escape their captors and set out to to save their friends, Blu and Jewel – two rare macaws and the stars of the Fox box office hit motion picture, Rio. Angry Birds Rio will pair the physics-based gameplay of the original game with unique twists based on the film.

Angry Birds Rio features:
- Five fantastic episodes with 135 exciting levels!
- Special Hidden fruits - discover them all!
- Put your Angry Birds skills to the ultimate test in a spectacular boss fight!

Expect plenty more where this came from: Episodic updates throughout 2011!

Customers who bought this also bought

Detailed overview of an app

Also, if you want to share your opinion about an app you downloaded, you can tap the Write a Review icon found in the upper right-hand corner of the review list. The Rate and Review page asks you to give an overall rating, a headline, and a review. Don't worry about space; there's room for up to a 3,500-character review. We're assuming you haven't used the app yet, so you can touch the Cancel button. When you're ready to write a real review, touch Post to upload it to the NOOK Store.

The Screen Shots area gives you an idea of what the app looks like before you download it. There are usually several shots so you can use your finger to scroll down and view them all.

Lastly, the More Like This section recommends apps that are similar to the featured app. For instance, if you're looking at a calculator app, NOOK Tablet might suggest apps related to finance or geometry.

It's fine if you decide against buying an app, too. Look in the upper right-hand corner and you'll notice an "X." Tap it to close the detailed app view and get back to your Apps Shop list.

However, if you do want to buy the app, tap the Price and Purchase icon. Whether it costs money or not, NOOK Tablet will ask you to confirm the purchase. You'll move through the following steps:

- Confirm
- Purchasing
- Downloading
- Installing
- Open

If your Wi-Fi is solid, the Purchase and Confirm steps will take just a few seconds. Downloading time depends on your Wi-Fi speed and

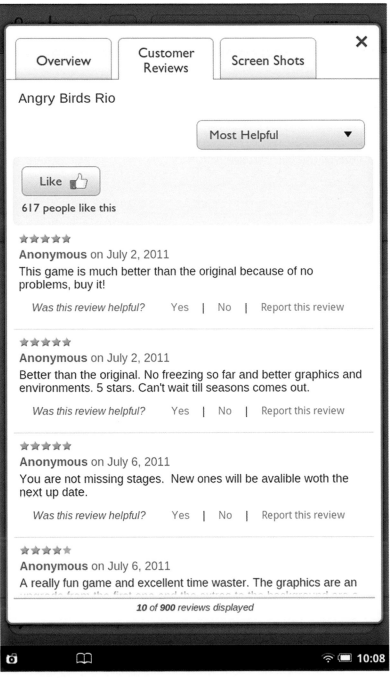

The rate and review app options

the size of the app. For instance, a simple calculator app could take a few seconds, while a robust visual app might take several minutes. As it downloads, a green bar will appear on top of the app icon. Your app is downloaded when the bar is full.

Next, NOOK Tablet will install your app. Again, the installation time will depend on the size of the app you downloaded. It should take no more than a couple minutes.

Once the app has finished downloading the button will read Open. Tap the button and your app will start up. When you're on the app's main page, you can just tap the icon to start up the app.

Managing Apps

As you explore more apps in the Apps Shop, you'll find several that will intrigue you and others that just won't be as interesting. Some, as we mentioned earlier, will be useful, but take up more memory than you'd like or are not needed frequently. Regardless of the reasons, you'll want to manage your apps.

First, go to your main Apps screen. Here you can tap an app to start, but try holding your finger down on an icon of an app you downloaded. You'll get the following menu:

- Open
- View Details
- Recommend
- Add to Home
- Add to Shelf
- Archive
- Delete

Open will start the app for you. View Details and Recommend will give you the info found on the app's Purchasing page and allow you to tell friends about the app.

The three most important options here are Add to Home, Add to Shelf, and Archive. Add to Home lets you move a link to the app on your NOOK Home Screen. Say you use a calculator app a lot, so you'd prefer not to have to start up your NOOK Tablet, pull up the menu, tap the Apps option, and finally tap the calculator icon every time to use it. If you choose to add the icon to your Home Screen, NOOK Tablet will create a quick link to your app. Turn on NOOK Tablet and you'll see the app sitting at the bottom of the screen. Remember that it is just another link to your app, so your app can still be found on your actual Apps screen, too.

Add to Shelf will move the app to a specific virtual shelf on your NOOK Tablet. It will ask you which shelf you'd like to move it to or the name of the new shelf you'd like to create. Now you can organize your apps using your own system.

However, if you do want to remove the app from your NOOK Tablet then the Archive option is perfect. As we touched on earlier, archiving takes an app off your NOOK Tablet and puts a link to it on the Archived page. Archiving an app can save you memory, because the item is removed from NOOK Tablet and it can keep your app page from getting too cluttered. If you'd like to get an app back onto your NOOK Tablet, tap the Archived icon that appears in the drop-down menu that opens after you tap "my stuff" in the Library, to see a list of all your archived apps. Choose the app you'd like to retrieve and tap the button labeled Unarchive. You can now get the app and go through the download process again. Keep in mind that NOOK Tablet will need Internet access to re-download your apps from the archive.

The final option, Delete, will permanently remove apps that you have no interest in downloading again.

NOOK Tablet comes with some games pre-installed. Did you check out the last chapter on apps? Your NOOK Tablet doesn't differentiate between apps and games, so your game collection will always be on your Apps screen.

To get to the Apps menu, tap the NOOK button on your device and open up the Quick Nav Bar. Touch the Apps icon. The Apps screen will now appear.

Top Apps

Not sure where to start? It can be pretty overwhelming to decide which apps you should download first. NOOK Tablet has a ton of choices, but certain apps are perennial favorites on the device. The apps are found in the Apps area of the NOOK Store. Here are some high-rated best sellers:

Bling My Screen • *Murtha Design*

NOOK Tablet has plenty of modification options, but extra choices are always welcome. Bling My Screen™ helps you personalize your Home Screen with additional wallpapers, virtual bookshelves, and other organizational tools.

Fandango Movies • *Fandango*

Like movies? The Fandango® app makes it easy to find out what movies are playing, learn about the films, and even purchase tickets for tonight's show. It also gives lots of theater-buff detail, from critics' reviews to box-office sales.

8. Games

A s we discussed, NOOK Tablet comes with a handful of pre-loaded games:

▶ Chess

▶ Sudoku

▶ Crossword

Chess and Crossword are digital updates of the classic board game and Sunday newspaper diversion, respectively. Sudoku is a popular number game.

Understanding Controls

NOOK Tablet doesn't have any physical buttons aside from the NOOK button. Its power lies in the flexibility of the touchscreen itself.

When it comes to games, your NOOK Tablet uses a few different control techniques:

▶ Tap

▶ Double tap

▶ Slide

▶ Hold

Tap is the method you've been using to do different things with your NOOK Tablet so far. It simply means touching the screen. You tap icons, books, and even menu choices to open them.

Double Tap means tapping the touchscreen twice in rapid succession. It isn't necessary for most actions, but you'll find that some games will require it.

Slide is touching a part of the screen and dragging your finger to another part of the touchscreen. A slide can move a game character

Chess

from one place to another, map the direction of a toss in a game, or several other actions.

Finally, Hold is doing a tap, and then keeping your finger down. It requires having a steady hand, as NOOK Tablet can sometimes think that you are sliding instead of holding.

It might help to see the touchscreen moves in action, so why don't we take a game for a test drive. Tap on the Chess icon.

If you know chess, everything will look familiar to you: 16 pieces on each side of an 8-by-8-square board and a timer to see how long each player is taking with his or her turn. Pretty basic, but how do you actually play chess on NOOK Tablet?

As you are learning, the touchscreen can create virtual, touchable buttons anywhere on it. Tap the New Game icon and a new game starts. Touch the Settings button and you can modify your color, difficulty level, and time limits. Once you get a new game started, try touching one of your pawns—one of the eight short pieces located in the second-to-bottom row on the board. Hold your finger on it, then drag your finger to the square above it. Look carefully and you'll see your pawn following your finger. Let go of the screen and, assuming it is a legal move, the pawn will move to the new square. You just slid a piece to where you wanted it to be and completed a move.

One more example might help here. Tap the NOOK button to get the Quick Nav Bar, touch the Apps icon, and tap the Crossword icon.

Again, the game looks like your traditional Sunday newspaper crossword: a big, blank graph with clues for the answers going across

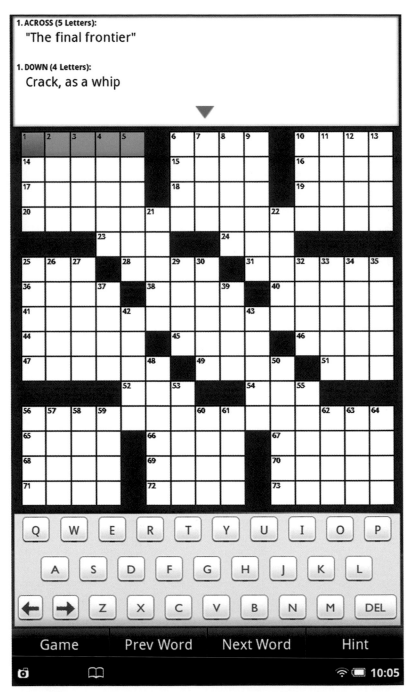

Crossword

Categories of apps for sale

and down. Tap on a square, and the clues for the corresponding across and down answers will appear at the top.

Notice how one line, either across or down, is highlighted? If you start typing in an answer, the game will automatically begin filling in that line. However, what if the line is going across and you want to go down, or vice versa? Double tap the square and the game will flip the highlight vertically or horizontally. In other words, tap once to put your focus on a particular square, but double tap to highlight either across or down.

Like apps, each game has its own particular set of controls. You can check the Help or Settings page of the game. Or, if you are feeling adventurous, play around with different touches and find out what they do. You can always reset the game!

The Games Screen

The last chapter described the Apps screen, which is where your games will be stored. To get to the Apps menu, just tap Apps in the Quick Nav Bar, and you'll be taken to the Apps screen in the Library.

Shop Now

Before you go on a shopping spree, check out the lower-left hand corner for the Sync icon (looks like two arrows making a circle). Sync lets NOOK Tablet check to see if there are any updates to your apps, including your games. Software can have bugs, like little glitches that make the game less enjoyable than it should be. And with games, sometimes companies will add free content just to keep you playing! Either way, game companies will give you a heads-up when a new, improved version of your game is out. Tap the Sync icon and your NOOK Tablet will download whatever updates are available for the

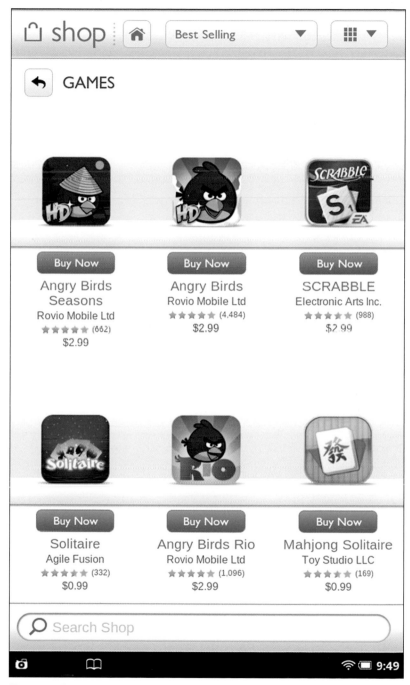

Shopping for games

games you own. Like archiving, syncing requires a Wi-Fi connection. Download times can vary, but, to be on the safe side, use the Sync option when you know you'll have more than a few minutes. In fact, your local Barnes & Noble store has free Wi-Fi, so you can always visit, sync up your NOOK Tablet, and browse the NOOK Store while the device downloads the latest versions of your favorite games.

The Apps Shop

Ready to get your hands on some new games? Like apps, your entry to games is through the Apps area of the NOOK Store.

Get back to the Apps Shop by touching the "n" button, selecting Shop, and then Apps. Similar to the books area in the NOOK Store, the Apps Shop shows the following parts:

- Order and Layout
- App Details
- Browse and Search Bar

The Order and Layout icons will change the format of the list of apps, Apps Details gives info on the highlighted app, and the Search Bar lets you type in a particular term to find the game you want.

Each game will have the Picture, Title, Company, Current Rating, and the Price and Purchase button.

Tap the Apps icon under the NOOK Store. It will give you a list of categories, from Children to Tools & Utilities. We want games, so tap on that category. There are thousands of apps, but now NOOK Tablet will only show you the games.

There are now three areas:

> Genre

> Bestselling

> Search

Genre will only show you games of certain categories, like Kids or Arcade. Bestselling lists the hottest games of the day.

Finally, searching is the most powerful tool you have in the Apps Shop for finding games. Tap the Search Bar and you'll get the following options:

> Search Suggestions/Recent Searches

> Search Bar

> Keyboard

Go ahead and type in the name of a game with the keyboard. Let's try the name of the super-popular game Angry Birds™. You'll notice that, as you type, the suggestion box will fill up with several different apps that fit the name. You can now tap one of the suggestions and open up that particular app in the Apps Shop.

Trying to remember something you searched for a while ago? Once you do a search, NOOK Tablet will remember the search terms you used. These are shown before the search suggestions pop up when you begin typing in a term. It is a quick, simple way to pick up where you left off during a previous search.

Downloading Games

Let's start downloading games. Find a game you like and, as with other apps, you'll get details, including:

> Name

> Company

- Version
- Overview
- Customer Reviews
- Rating
- Screen Shots
- Price & Purchase
- Add to Wishlist
- Share
- More Like this

Many of these details were discussed in the last chapter, but the most important ones are Price, Customer Reviews, and Screen Shots, so you can see how the game looks. The recommendations section, which tells you "Customers who bought this also bought..." can also be helpful after you buy the game because, if you like it, you can find out about other similar apps.

If you decide not to buy a game, you can touch the "X" in the upper right-hand corner. This will bring you back to the general games list.

When you find a game you would like to buy, touch the Purchase icon. Your NOOK Tablet will ask you to confirm the purchase. To approve it, you'll use the following process:

- Confirm
- Purchasing
- Downloading
- Installing
- Open

Confirm and Purchasing should just take a few moments. Downloading a game takes a little while longer, depending on your Internet connection and the memory size of the game. Big, complicated games take longer to download. You'll see a green bar on top of the Games icon that tracks how much has downloaded so far. Stay in your Wi-Fi area while the game is downloading, or the process may be interrupted.

Once it finishes downloading, the game will automatically install on your NOOK Tablet. Finally, when the Open button appears, you're ready to play! Tap it and the game will start. The game is also installed on your main Apps screen. You can always tap the icon from there, too.

Managing Games

As was mentioned earlier, some games will take up a lot of memory in your NOOK Tablet. In fact, games in general tend to be bigger in size, so the chances of you wanting to archive the latest NOOK Tablet game to save on memory are a lot higher than, say, needing to archive a simple calculator app. As a reminder, archiving means taking an app off your NOOK Tablet but keeping it available, via Wi-Fi, for re-download whenever you'd like to play it again.

To organize your games, go to the main Apps screen. Hold your finger on any game icon. You'll see a menu pop up:

- Open
- View Details
- Recommend
- Add to Home
- Add to Shelf
- Archive
- Delete

Open, View Details, and Recommend are fairly straightforward. Add to Home lets you put the icon on your NOOK Tablet Home Screen. Play one game a lot? You may want to put it on your Home Screen so you have the quickest access to it, whenever you want to play your game.

Archiving is the key to keeping your game (and app) collection tidy and orderly. While you're on this menu screen, tap Archive and NOOK Tablet will take the game off your device and move it into online storage. Whenever you want the game back on your NOOK Tablet, tap the Archived icon, which appears in the drop-down menu that opens after you tap My Stuff in the Library, and you'll get a list of all the apps you've archived. Remember that you need a solid Internet connection to unarchive any games. You can also permanently remove a game by selecting Delete.

Top Games

NOOK Tablet has a serious collection of games for both joystick jockeys and casual gamers. Here are some favorites:

Angry Birds • *Rovio*

One of the most popular mobile games ever made, Angry Birds has you tossing birds to take down the structures of the greedy pigs. Underneath the cartoon atmosphere is a super challenging (and addictive) puzzle game that teaches physics.

Solitaire • *Agile Fusion*

A solid version of the classic card game, Solitaire pushes you to put random cards in order as quickly and as smartly as possible. Peppered with relaxing music and nice graphics, Solitaire has different game modes to keep things challenging.

Doodle Jump Deluxe • *GameHouse*

Doodle Jump® was one of the first major hit tablet games, and NOOK Tablet's update keeps the spirit of the original. Simply slide your finger or tilt the device to bounce from platform to platform. The higher you go, the bigger the points and the more worlds you can explore.

Scrabble • *Electronic Arts*

The word battle Scrabble® has already been a classic in board game form, but NOOK Tablet's version adds plenty of new features. The hint system and tough computer opponent are great. The biggest benefit, though, is being able to play against your friends online competing for the best word score.

9. Productivity

W e've discussed fun apps and gaming in the previous chapters, but what if you actually want to get work done on your NOOK Tablet? The virtual keyboard, copious memory, and easy portability make that easy to do, using the dozens of productivity apps in the NOOK Store.

Like other apps, all your productivity software is on the Apps screen in the Library. To get to the App menu, touch the "n" button to pull up the Quick Nav Bar, tap Shop, and then the Apps icon. In the Genre box at the top, scroll the catagories until you see Productivity.

To find a certain Productivity app, you'll want to use the Search feature. Tap the Search Bar and you'll find the following options:

- ❱ Recent Searches
- ❱ Search Bar
- ❱ Keyboard

Using the keyboard, type in "Quick Office," the name of the popular word-processing suite. You'll notice that, as you type, the suggestion box will fill up with several different apps that fit the name. You can now tap one of the suggestions and open up that particular app from the NOOK Store.

Trying to remember something you searched for a while ago? NOOK Tablet will remember the search terms you used. These are shown before the search suggestions pop up when you begin typing in a term. It is a quick, simple way to pick up where you left off during a previous search.

Productivity apps

Downloading Productivity Apps

Like all apps, your office apps will give you the following:

- Name
- Company
- Version
- Rating
- Price and Purchase icon
- Add to Wishlist
- Customer Reviews
- Share
- Screen Shots
- Overview
- More Like This

Pay attention to the price, the reviews, and the screen shots before you purchase. If you don't want to buy an app, tap the "X" in the upper right hand corner to get back to the main apps list.

When you find a productivity app you want to buy, tap the Buy Now button to begin the process:

- Confirm
- Purchasing
- Downloading
- Installing
- Open

Confirm and Purchasing are quick, but downloading the actual program can take a few minutes. Watch the green bar that appears on the app icon to see how long you have to wait. Remain in a Wi-Fi hotspot to make sure the download isn't interrupted.

After it downloads, your NOOK Tablet will install the app. The Open button appears once the app is ready. Tap it to start your app. Remember, you can always run your app from Apps screen in the Library, too, as your icon will appear there.

Managing Apps

Check out Chapter 7, *Apps* for more details on organizing apps, but know that holding your finger on the Apps icon will give you a menu with the following options:

- Open
- View Details
- Recommend
- Add to Home
- Add to Shelf
- Archive
- Delete

Top Productivity Apps

There are many apps that will help transform your NOOK Tablet into a full-blown tablet. Check out this software to get work done on the go:

Quickoffice Pro • *Quickoffice*

Quickoffice® Pro lets you stay on top of work commitments while you're on the go. The office suite is compatible with Microsoft Office™ Word, Excel, and PowerPoint, and it syncs up to your documents through Google Docs, Dropbox, and other software.

Calculator & TipCalc • *5ivedom*

This low-priced app features a built-in tip calculator. The main calculator, capable of sine, cos, and tan measurements, isn't too bad, either.

My-Cast Weather Radar • *Garmin Digital Cyclone*

NOOK Tablet doesn't come with a weather app, and this handy one from the GPS leader Garmin fits the bill. Download My-Cast® for hourly or extended forecasts, detailed maps and graphs, or international weather information for your travels.

10. Getting Social

Not only is NOOK Tablet light and portable, but the Internet connection makes it easier to make new friends online, reconnect with old friends, or even share your books with other people. All this can be found in NOOK Friends.

To start up NOOK Friends, press the NOOK button to open the Quick Nav Bar. Tap Apps and then NOOK Friends on the Apps screen.

There are three icons under the NOOK Friends app:

- Friends' Activities
- About Me
- LendMe

The default page, Friends' Activities, gives you a news feed telling you what your NOOK Friends are doing. Your friend Sarah might have just recommended the latest Mary Roach book, while Paul just reviewed his latest sci-fi read. It's a good, simple way to stay connected to your fellow readers. You can look at it as a virtual book club.

NOOK Friends lists all your current reading-circle friends. You can also request friendships and send messages to current friends. We'll get into that in a moment.

The About Me page is the personal profile people see when they are your friends. It has a few facts about you, including the number of friends you have and the books you currently own.

Finally, LendMe tracks all the books that you've lent and borrowed, as well as titles you've asked to borrow and books people are willing to lend.

LendMe Library

If you own any books that you don't want others to see, there is a way to hide these titles from view. Learn more about this in Chapter 12, *Advanced Techniques*.

Finding and Making NOOK Friends

Tap on the All Friends icon to see all your NOOK Friends. You'll see three areas:

- ▶ Add Friend
- ▶ All Friends
- ▶ Pending

Let's actually start with the second icon, All Friends. Tap the All Friends button and you'll see a list of your current friends with their name, picture, and other stats. You haven't connected with anyone yet, so this section will be blank. The next section, Pending, shows all the people interested in being friends with you and, conversely, all the folks you've asked to be your friend but haven't responded yet. If someone has asked to be your friend, you can tap Accept (to become his or her friend) or Ignore (to make the request disappear). Don't worry: Like other social networks, NOOK Friends won't send a notice to people you choose to ignore. No need to feel guilty!

You can also drop NOOK Friends, too, by tapping the "x" icon just to the right of their name. Finally, you can borrow a book from them by tapping the LendMe icon and finding out what books they have available to lend to you.

NOOK Friends

Now, let's start getting some friends. Tap the Add Friend icon and NOOK Tablet will give you four ways to find friends:

- Find Friend from My Contacts
- Find Friends from Facebook
- Find Friends from Google
- Invite a Friend via Email

If you already know one of your friends is a NOOK user registered on BN.com (Barnes & Noble's Web site), you can tap Add New to contact him or her directly. NOOK Tablet will ask for the first name, last name, and email. Barnes & Noble will send your friend an email.

Suggested Contact is cool because you can also contact friends who don't have a NOOK yet, but may get one in the future. You will still go through the Add New process, but your NOOK Tablet will store the person's email address and give you a heads-up when he or she registers on the BN.com Web site. Your contacts who are newly registered BN.com members will be found under the Suggested icon.

Finally, you can have NOOK Tablet go through your social media contacts and see if any of your friends are NOOK users registered on BN.com. Tap the All Contacts icon, then the Set Up Account icon. You'll find three different ways to find new NOOK Friends:

- Facebook
- Twitter
- Gmail

You can choose the social media where most of your friends are listed. For instance, if you're a hard-core Facebook user but never touch Twitter, you can just link your Facebook account. There may be more social media options in the future, but Facebook, Twitter, and

Gmail cover literally millions upon millions of users, all of whom are potential NOOK Friends.

Facebook is the easiest way to make connections because, once you sign in, will automatically make you NOOK Friends with your Facebook friends. On the other hand, Twitter won't allow NOOK Tablet to search your friends lists for NOOK users, but it does make it easy for you to post any new book, app, or review onto that social network.

Connecting your NOOK Tablet to Facebook or Twitter is the same process: Tap Link Your Account, then type in your email and password. Your NOOK Tablet will confirm that you're allowing it to post onto your Facebook wall or Twitter feed. Now you can update friends about your NOOK Tablet activity and let them know that they can "friend" you through NOOK Friends.

If you have a Google account, and particularly a Gmail address, you can have your NOOK Tablet search your contacts and find any current friends on NOOK. (Luckily, Gmail is free. You can get your own account at www.gmail.com.) Like Twitter and Facebook, Gmail will ask for your email, password, and permission to connect to NOOK. Now any current registered NOOK users on your Gmail contact list will appear in the Suggested area. Because NOOK Tablet has to sort through the contacts, they usually take a few minutes to appear on the list.

Lending / Borrowing Through LendMe

Sharing good books with friends is one of the pleasures of physical books; you can lend a worn paperback to a friend and she can return it to you when she's done. If your friend happens to lose the book, it can be easily replaced.

You can't just hand off your NOOK Tablet to a friend so he can read one book; you'd be handing off your whole Library! Also, replacing a lost or damaged NOOK Tablet costs a bit more than the price of a paperback.

Luckily, your NOOK Tablet makes it easy to share your books with other NOOK users. They can borrow as many books as they like, and you, in turn, can borrow books from them, too.

The NOOK lending and borrowing program is called LendMe. To see your LendMe section, tap the LendMe icon on your NOOK Friends page. It is the third button.

LendMe has four sections:
- **Lend:** my lendable books
- **Borrow:** friends' books to borrow
- **Offers:** books friends will lend
- **Requests:** books others want to borrow

Lend shows the books in your collection that you can let others borrow. Like a physical book, your NOOK books can only be lent out to one person at a time and, on the NOOK Tablet, you can only lend a book one time. Your friend has a week to take you up on your lending offer—otherwise the book comes back to you. If they do decide to borrow your book, they keep it for 14 days, a period during which you won't be able to read it. Keep in mind that some publishers and authors don't make their books available for lending, so NOOK Tablet will only show those books that can be shared.

Borrow shows all the books your friends have that you can borrow. You can request a book to borrow here.

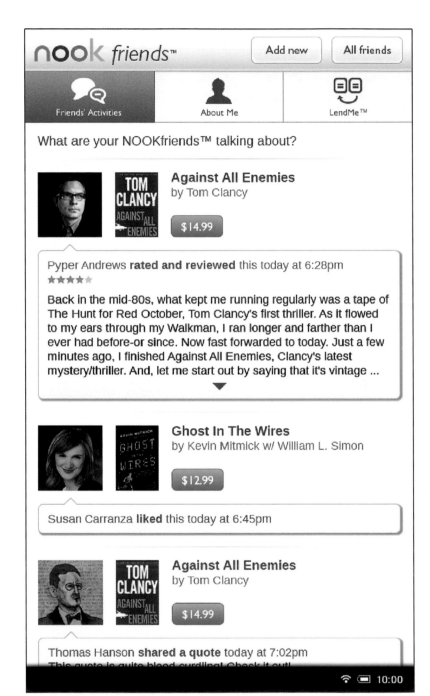

Comments from your NOOK Friends

Offers are books that your friends say they would be happy to let you borrow. This can come in handy. Let's say you are having coffee with a friend and she mentions a book she recently read that you'd absolutely love, but she can't remember the title. Once she gets to her NOOK, she can look up the book and offer it to you to borrow. Finally, Requests are books your friends would like to borrow from your current Library. You can let them borrow your book from here.

> Want to keep some of your reading material private? In the NOOK Friends section, tap the Privacy button in the upper right-hand corner. You'll now see a list of all your books. Toggle the Show icon to make a book visible or invisible to your friends.

Using Email

Another useful feature of your NOOK Tablet is email. It is located on your Apps screen, which you can reach by pressing the NOOK button.

NOOK Tablet will ask you for your email address and password, the type of email (POP or IMAP), and other details specific to your email. If you're not sure how to fill out the details, go with the default information. If it doesn't work, visit the Help page of your email Web site, and it should tell you how to allow devices to have access to your email.

Once your email is set, you're ready to access your messages. Keep in mind that NOOK Tablet doesn't create a new email account for you, but just allows you to read and send email through your current accounts, like Gmail or Yahoo.

11. Memory and Storage

All digital gadgets have a limited amount of memory, and over time the memory gets used up. NOOK Tablet has space for thousands of books but the Library can get cluttered, especially when you start downloading apps, music, magazines, and newspapers. It's worth knowing how much memory you have, and how you can help mitigate digital overflow.

In order to see how much space is available for more books, periodicals, and apps, access the Settings menu by pressing the NOOK button to open to Quick Nav Bar. Then tap Settings. Once there, tap the tab that says Device Info.

The Device Info pane shows how much battery power is left, and how much memory is left. You can go to your Advanced Settings and look under Device to see how much memory you have available. That space can hold countless books, apps, or anything else, but it's important to know how to manage and delete items just in case NOOK Tablet begins to run out of storage space.

In order to see how large a specific book is, simply double tap that book in the Library. The book will tell you where it is stored and how large it is. The page will also give you information such as the title, author, and when the file was last modified.

Managing Space

Now that you know how much storage space is left, it's time to learn how to manage it. There are three different places to store books:

- In NOOK Tablet's memory
- On the Barnes & Noble Archive online
- On a microSD™ Card

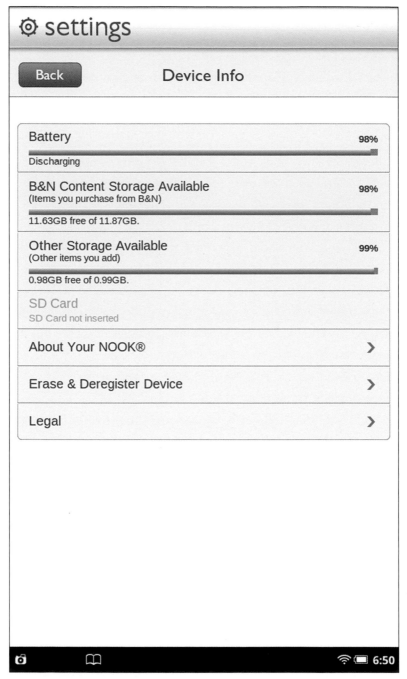

NOOK Tablet's Device Info screen

Archiving Items

To archive an item, go to the book's detail page by tapping and holding the book's cover in the Library, and tapping Archive. You should receive a confirmation message asking if you really want to do this. Tap Yes, and the book will be removed from your NOOK Tablet and stored in your account on Barnes & Noble's Web site. You can find it under B&N Lifetime Library™.

You can only archive books you've purchased from the Barnes & Noble store. Anything that you've put on your NOOK Tablet from your computer can be added and removed at your own discretion. We'll be discussing how to connect your NOOK Tablet to your computer later in this chapter.

The Barnes & Noble Archive

The Barnes & Noble Archive is a list of books that aren't currently on your NOOK Tablet but are available for you to re-download. To access the Archive, go to BN.com and log into your account. Once there, move your mouse over the My Account tab and click My Library. On the left pane, there will be a button to view your Archive. In the Archive, you can delete anything you no longer want. The Barnes & Noble Library also lists what samples you have downloaded, and what books have been purchased in full.

Re-downloading Items

Once your item is archived, it's extraordinarily easy to get a book back onto your NOOK Tablet. To do so, make sure you have a Wi-Fi connection. Then, go to the My Stuff section and tap the Archived option. Tap the book you want to unarchive, and the book starts to download back onto your NOOK Tablet. It's a one-tap process, and the book should be in your Library in a matter of seconds.

MicroSD Cards

The final way to manage space on your NOOK Tablet is with microSD cards, which are sold at most electronics stores and can be used for storing an additional 32 GB of books and personal files on NOOK Tablet. In order to use it, you first need to install the SD card. To do so, turn off the device and insert the card in the slot by your NOOK Tablet's curved metal bar.

Now that the card is installed, you can load files onto it from your Library. In Library, go to the My Files tab in the upper left corner, tap it, and look for the button labeled Memory Card. Here, you'll be able to find all the PDF and eReader files on the card that NOOK Tablet can read.

If you want to transfer files to NOOK Tablet you'll have to connect the NOOK Tablet to your computer, which we'll explain now.

Connecting to a Computer

Connecting NOOK Tablet to a computer is easy. Just take the USB-to-micro-USB cable that came with the device, plug the micro-USB end into NOOK Tablet and the USB end into the computer. Now the two are connected.

From here, you can manually add files to NOOK Tablet. To do so, simply right-click and select Copy for the files you want to put on the NOOK Tablet, navigate through the file named My NOOK and paste them in. It doesn't matter what subfiles they're in, as long as they're in a supported format and are stored on NOOK Tablet. From there, the files should be accessible from the Library.

> For this section, we're talking exclusively about how to connect NOOK Tablet to a PC or Mac. The process could be different for a Linux system.

By connecting your NOOK Tablet to a computer, you'll be able to add free eBooks from Web sites, such as the Gutenberg Project, which we'll be discussing later in Online Resources in the Appendix.

eBook files that NOOK Tablet can read:

▶ ePub

▶ PDF

Wishlists

If you want to make a note to purchase something later, then Wishlists are the way to go. On the page for every book, right next to the book's star rating, there's a little box labeled My Wishlist. Tap the box to put a check mark in it. Every book you check mark will be on your Wishlist. The Wishlist is accessible by tapping the My Account button in the upper right corner of the NOOK Store. This opens a pull-down menu. Tap My Wishlist. With the Wishlist, you shouldn't have any trouble remembering which books to check out.

12. Advanced Techniques

NOOK Tablet is filled with outstanding features and capabilities, many of which take you well beyond the simple enjoyment of reading. There are ways to create a personalized wallpaper for your NOOK Tablet, plug your device into your computer to access eBooks that were not purchased from the NOOK Store, manage an address book, and connect to several social networks, like Facebook and Gmail. There's also Barnes & Noble's LendMe Program, which we talked about in Chapter 10, *Getting Social*. You can also hide books from the LendMe list, which we'll talk about here.

We're also going to get really technical in this chapter. Here's where you'll find out how to use NOOK Tablet outside of the United States, and how to update the entire device with the latest software.

Taking Advantage of Barnes & Noble Stores

There are unique benefits to using your NOOK in Barnes & Noble stores.

Bring your NOOK Tablet into a Barnes & Noble store, and you can do several things:

- Use Wi-Fi for free.
- Read books for free for up to one hour per day.
- Get technical support for your NOOK Tablet.
- Return a device for repair.
- Download exclusive content.
- Get a personalized recommendation for a new book.
- Try out new NOOKs.

Perhaps the biggest benefit is being able to use Wi-Fi for free!

Through a special arrangement with AT&T, all Barnes & Noble stores have free Wi-Fi. Let's say you don't have a local wireless connection at your home or at any nearby coffee shops. You can just come to your nearest Barnes & Noble and download whatever books, periodicals, or apps you've had your eye on.

You can also read books for free. As we talked about earlier in the book, you can always download samples of books, periodicals, or apps. However, if you bring your NOOK Tablet into your local Barnes & Noble, you can also download entire books for reading. Barnes & Noble gives you a full hour with a book of your choosing. Your NOOK Tablet will let you know when the time is up.

> Barnes & Noble gives you an hour a day per book to read for free, so you can always come back the next day to read more!

Barnes & Noble booksellers are known for being voracious readers as well as hearty app users, so visiting a local store is an excellent way to find out about the latest and best content for your NOOK. Not sure if the newest book from your favorite novelist is worth reading? Trying to decide on the best international newspaper to read for an upcoming trip? Come into the store and chat with a Barnes & Noble bookseller; they'll help get you going in the right direction. And while this book hopes to answer all your NOOK questions, sometimes it helps to have someone give you hands-on assistance. The folks at Barnes & Noble stores know all about your device and are happy to help answer your technical questions.

Barnes & Noble stores also offer a faster way to get your device repaired. If you chat with a Barnes & Noble bookseller and still aren't able to get your NOOK Tablet up and running, the bookseller may suggest sending the device in for repair. Before you do, you'll want to ask the bookseller:

▶ What the problem seems to be

▶ If the problem falls under warranty

You might also want to read over the FAQ as well as the warranty information (basic and extended) that are listed at the very end of this book. If the problem doesn't fall under warranty or if the warranty has expired, Barnes & Noble won't be able to repair your device. It's important that you know these answers before you hand it off to the bookseller to ship in for repair. However, if your warranty is current and the problem falls under the warranty, the assistant will be happy to send your NOOK Tablet in for repair.

Come into a Barnes & Noble store and you can also download exclusive content. There are thousands of books, periodicals, and apps in the NOOK Store, and dozens more come in every day. But Barnes & Noble also offers interesting, exclusive goodies that can only be downloaded at the store. No need to plug in your NOOK Tablet—as long as you are in the store, the local Wi-Fi will give you access to the content. You can ask the Barnes & Noble booksellers what exclusive content is available at the moment.

You can also take the new NOOKs for a test drive. If you have a NOOK Tablet, why not check out the lower-priced NOOK Color or the light NOOK Simple Touch? And, if you have a NOOK, you can play with the powerful, versatile NOOK Tablet. It's a fun way to learn all about the latest and greatest capabilities offered by eReading.

Background Images

Tired of the picture in the background? You can switch up the wallpaper anytime. Go to the Home Screen and hold your finger on any empty part of the screen—in other words, directly onto the wallpaper. You'll get the option to Change Wallpaper.

There are three ways to change the background:

- Wallpaper
- Photo Gallery
- Live Wallpapers

Wallpapers are static pictures that sit in the background as you use your NOOK Tablet. Your device comes stocked with a handful of wallpapers that you can choose from.

Photo Gallery has photos that you have downloaded, either from someone else or from your own personal collection. Your NOOK Tablet also comes with a few sample pictures.

Finally, Live Wallpapers are cool animated pictures that move in the background as you use your NOOK Tablet. These are available from retailers in the NOOK Store.

If you'd like to upload your own photos, plug your NOOK Tablet into a computer, and wait for the machine to read the device. Once you're connected, go to the file marked My Computer and double click on the NOOK folder. Now you are looking into your NOOK Tablet memory.

Depending on what you want to add, click on the file marked Wallpapers or Pictures. Here, create a new folder with a one-word file name. Double click on it, and add any image files you want. This is best done by right-clicking Copy on the image, and by right-clicking Paste in the file you've created.

Image files the NOOK Tablet supports:

- JPEG
- GIF
- PNG
- BMP

Reading PDFs and Other Documents

NOOK Tablet isn't restricted to only reading eBooks and other eReader files. The device can also read PDF files. The process of adding them onto NOOK Tablet is easy, and very similar to adding ePub files (as detailed in Chapter 11, *Memory and Storage*).

In order to load PDF files to your NOOK, start by plugging your NOOK Tablet into a computer. Next, right-click and copy the PDF file you want to add, and paste it into the NOOK file labeled My Files. After that, safely disconnect NOOK Tablet, push the Quick Nav button, and tap Library. From there, navigate through the folder tab on the top left corner, locate and tap My Files. You should then be able to locate all of the PDF and ePub files that weren't downloaded through the NOOK Store. The PDF files are marked with an icon featuring a PDF logo.

Once you've found your PDF file, double click on it, and it should open up just like an eBook. (Note that you can't search for text in PDF files saved as images.) In fact, the layout is exactly the same as an eBook. Simply read the PDF file the same way you read an eBook, and there should be no problems. If there are, however, consult Chapter 13, *Troubleshooting*.

Updating Your NOOK Tablet Software

Keeping your NOOK Tablet updated with the latest operating software is both simple and important to do. It allows you to benefit from any changes Barnes & Noble has made to make NOOK Tablet a better product, and to keep everything running smoothly.

NOOK Tablet actually updates on its own. When you're connected to Wi-Fi, it will automatically download any updates without any input from you. When an update has been installed, a new button will appear on the lower left of the screen, right next to the button that takes you back to your most recent book. Tapping the button will let you know that a new version of NOOK Tablet software was successfully installed.

If you want to know what software version your NOOK Tablet is using—this is important sometimes for troubleshooting and discussing new features that may be added to NOOK Tablet—the process is simple. Push the NOOK button, tap Settings, then tap Device Info. From there, tap the button labeled About Your NOOK. The software version should be listed in the middle of the table, right under your account address and the model number.

13.
Troubleshooting

NOOK Tablet is a user-friendly device, but it can occasionally run into some technical problems. The following are a few fixes that should help you solve potential problems.

Not Charging?

If NOOK Tablet isn't charging, make sure the device is plugged in. You'll know it is being charged when the orange light at the bottom of the device is lit up. When the device is fully charged, the light turns green.

Not Connecting to Wi-Fi?

If you're having trouble connecting to Wi-Fi, make sure that your NOOK Tablet is looking for the desired Wi-Fi. Push the NOOK button, press Settings and tap Wireless. From there, make sure the Wi-Fi is turned on and is connected. If not, either connect to an alternative Wi-Fi network, or troubleshoot your wireless router.

Music or Video Isn't Playing?

Your NOOK Tablet is a versatile device, but it can only understand certain music and video files. As far as music, it can play AAC, amr, mid, MIDI, MP3, M4a, wav, and Ogg Vorbis files. With video, it plays Adobe Flash, 3gp, 3g2, mkv, mp4, m4v, MPEG-4, H.263, and H.264 formats. Other video types aren't guaranteed to work.

What Kind of Customer Support Does Barnes & Noble Offer?

Barnes & Noble understands that NOOK Tablet is an investment on your part, so the company has provided several ways for you to get the technical support you need:

- Phone
- Email
- Live online chat
- In-store
- NOOK Web site FAQs

If you'd like to talk with a Barnes & Noble representative over your landline or cell phone, give the company a call at 1-800-843-2665. If you prefer email, the Barnes & Noble troubleshooting email is nook@barnesandnoble.com.

You can also talk with someone online, which is like using an instant messenger program to ask an expert your pressing questions. To use the live online chat, go to http://www. barnesandnoble.com/nook/ support/ and click on the Chat Now link under Chat With a NOOK Tablet expert. The Web site will ask you for five pieces of information:

- Name
- Email address
- Product (NOOK Simple Touch, NOOK Color, or NOOK Tablet)
- Serial or order number
- Question

Give your full name and the account email address that is associated with your Barnes & Noble account. Be sure to choose the right product, as the troubleshooting for your NOOK, NOOK Tablet, and NOOK Color are different.

Also, it helps if you have your serial number. If your NOOK Tablet is functional, tap the NOOK button to bring up the Quick Nav Bar, and then tap the Settings icon. Next, tap the Settings icon, touch Device Info, and then About Your NOOK. Here you'll find lots of details about your device. The serial number will be the second to last number listed. It will be 16 digits.

Finally, describe your question as thoroughly as possible. You'll be talking to someone—that's the whole purpose, of course!—but asking a clear, detailed question will make the process faster and easier. The more info the customer service representative has, the better they can help.

When you're ready, tap the Submit button. A Barnes & Noble representative will then hop online and answer your question.

And, if you'd like to talk to someone face to face, you can go online at barnesandnoble.com and find a nearby Barnes & Noble store. The knowledgable associates will be happy to help.

Other Issues

If your problem is not pressing, you can also use the Barnes & Noble forums. In fact, there's always an active community on the forums, so your question may have already been asked by someone else and answered by Barnes & Noble. You can check out the troubleshooting forums at: http://bookclubs.barnesandnoble.com.

14. Accessories

With all that you'll be using your NOOK Tablet for, you'll want to make sure that it looks great (and is protected) while you're carrying it around town with you. Barnes & Noble offers many options to protect and accessorize your NOOK Tablet, including cases, bags, and covers that will help your NOOK Tablet stand out.

Screen Shields

Barnes & Noble sells an antiglare screen shield that can be simply applied on top of NOOK Tablet. This cover helps protect the screen from dust and foreign particles while the screen maintains all of its touch functionality.

Cases

NOOK Tablet also can be put into a selection of cases. These simply slide over the device to add some color, protection, and style. Some can open from left to right like a book, while others don't have a spine at all. They come in leather and cloth and include designs from famous designers like Jonathan Adler and Kate Spade.

Other

Barnes & Noble also sells a stand for NOOK Tablet. This lets readers use the device and keep it at an optimal reading angle without needing to hold it.

Barnes & Noble sells car chargers for your NOOK Tablet, as well as extra AC adapters, to ensure the device can get power at any and all times.

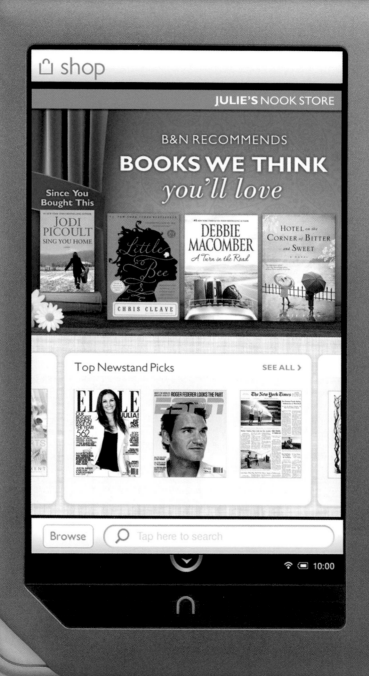

NOOK
Color

15. A Look at NOOK Color

In the Box

Inside the NOOK Color box you'll find everything you need to get started. The start-up process will take you only a few minutes, and, aside from optional accessories, you don't need to make any other purchases. We'll get into NOOK Color covers and other fun additions later in the book, but for now let's crack open the box and see what's packed inside.

First, take off the wrapper. Then, bend the bottom half of the box to open the package. Here's what you'll see:

- ▶ Your NOOK Color
- ▶ A USB-to-micro-USB cord
- ▶ A wall plug

Don't see the cord and the wall plug? NOOK Color is at the top of the box, while the cord and the wall plug are snug inside the bottom of the container. Remove your NOOK Color and its accessories from the box.

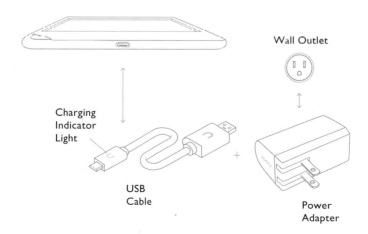

NOOK Color accessories

NOOK Color

NOOK Color is less an eReader and more like a full-fledged tablet. It can play Flash video; run apps (including video and word games) and use interactive, pop-up style books. The memory can also be expanded so you can download as many apps, music, and reading material as you can handle. You'll notice that the dimensions of NOOK Color are hefty compared to other eReaders. It is actually intended to go beyond just reading books, so the plush color screen and solid weight make sense when you're watching videos.

The front of your NOOK Color displays the full-color touchscreen. Right below the touchscreen is the "n" symbol, or NOOK button. One of the few buttons on NOOK Color is the "n" key found on the center-bottom of the device's front. Tap it when the power is on and your current options will pop up onscreen. Notice the curved hole in the lower left-hand corner? Please do not use it to hold your NOOK Color as it is a design element and is not reinforced.

Now tilt NOOK Color away from you so you can see the thin bottom right below the "n" key. Here is the hole for your USB wire. We'll plug it in after we finish our tour of the device. Turn the device to the right, past the hook, and check out the left-hand side. The lone button there is the Power switch.

Turn the device to the right again and you'll be looking at the top of NOOK Color. Here you'll find a single hole for your headphones. NOOK Color has a speaker for playing music or other audio, but the headphone jack means you can enjoy your audio on your own as well. The headphone jack is 3.5 mm—the industry standard—so you can plug virtually any mainstream headphone into the device.

Turn the device to the right one last time and you will see the left-hand side. See the plus and minus buttons? These will control your volume. Once we turn on your NOOK Color, a little volume meter will appear onscreen. If you do decide to use headphones, you'll want to check the volume before you put them in your ears!

Now flip NOOK Color so the touchscreen is facing down. Check out the textured back. While other devices are smooth and, arguably, slippery, NOOK Color's raised back makes it easier to keep a grip while you read, watch, or listen. And, if you look very carefully, you'll see several rows of holes near the bottom edge. These holes are the speaker. It may seem small, but the speaker packs enough punch for your multimedia listening.

Finally, let's take a look at the microSD card. Keep your NOOK Color facedown and give the metal NOOK Color logo by the hook in the lower right-hand corner a pull. The hidden SD card compartment will pop up.

You'll be downloading lots of books, periodicals and, perhaps, apps for your NOOK Color, all of which use a lot of memory. The microSD card allows you to expand the memory your NOOK Color can handle. We'll discuss microSD cards and download management further in Chapter 25, *Memory and Storage*.

Before You Start: Charge It

Okay, now that the tour is done, you'd probably like to start reading! It's hard to resist the urge to dive right in when you get a new device, but there is usually a step or two you have to take to guarantee a good experience. For NOOK Color, Barnes & Noble recommends that you charge up your device before you start playing with it.

Let's get it charged before we take it for a spin. Take a look at the USB cord. One end has a traditional USB plug or connector that connects to most computers, while the other end has a micro-USB format that is usually used by devices. Plug the wider USB end into NOOK Color's wall plug.

The micro-USB end plugs into your NOOK Color. As you may remember, if you look at the "n" button, you'll find the micro-USB hole right below it. Connect the micro-USB in there and, finally, put your wall plug into an outlet. If everything is connected properly, you'll see a small orange light coming from the micro-USB end. The light will turn green once NOOK Color is fully charged.

Powering Up

Did the light on the micro-USB change from orange to green? Your NOOK Color is charged. We can finally turn it on! Hold the Power key, located on the left side of the device. You should get a brief loading screen and then the intro page.

There are four steps to getting started here:

- Agree to Barnes & Noble Terms and Conditions.
- Set time zone.
- Connect to Wi-Fi.
- Register your device and default credit card.

You can also click on the intro video to get a brief personalized look at your device.

For now, let's go through the intro steps. The Barnes & Noble Terms and Conditions is the list of things you promise to do (and not do!) with the device. When you are done reading, tap on the Agree icon.

NOOK Color Home Screen

Wi-Fi

The next big step is setting up your Wi-Fi connection. A Wi-Fi connection is required to finish setting up your NOOK Color. Are you in a Wi-Fi hotspot? A hotspot is any area where a Wi-Fi router is in range. Wi-Fi routers at public venues are usually free to join, but others, like the Wi-Fi router in your house, usually are (and should be) password-protected. It's unlawful to use someone's personal Wi-Fi without his or her permission.

The good news is that every Barnes & Noble store offers free Wi-Fi. It is a great alternative if you don't have your own Wi-Fi router at home.

If you are in a Barnes & Noble store, your NOOK Color will automatically detect the store's Wi-Fi network and ask if you want to connect to it. When you are in a hotspot, NOOK Color will ask you to choose a Wi-Fi connection. If you are at home, choose your personal Wi-Fi connection name from the list. If the Wi-Fi router requires a password, NOOK Color will ask you to type it in. A virtual keyboard will pop up onscreen that will let you punch in the letters and numbers in the password.

If you are having trouble with your home Wi-Fi showing up on the list of hotspots, try unplugging your router from the outlet and plugging it back in after five minutes. Check the list of NOOK Color hotspots again and it should be on the list.

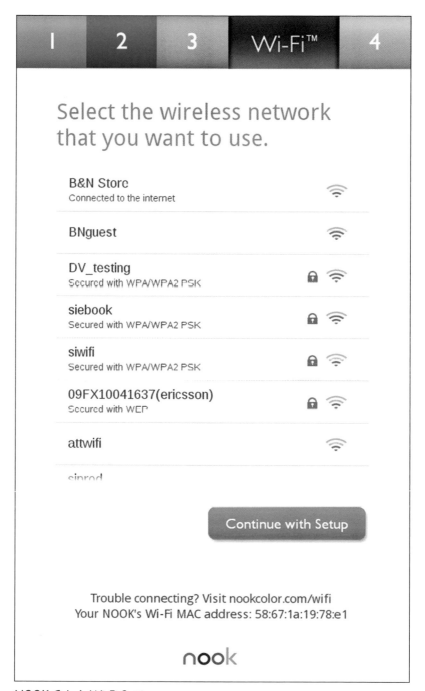

NOOK Color's Wi-Fi Settings screen

Account Information

The last step is registering your device, your account, and your credit card. Do you already have a barnesandnoble.com account? If you remember your username and password, type it in here. If you have an account, but don't remember the password, you can hop online and request a new password to be emailed to you.

If you don't have an account yet, go ahead and type in a username and a password that you will remember but others will have a hard time guessing. Barnes & Noble will check to see if anyone else is using the same username. As soon as you have an original username, you'll move on to the credit card.

Like other mobile devices, your NOOK Color allows you to make purchases without having to type in your credit card information each time. Instead, you type in a default credit card (it must be a credit card from a U.S.-based bank, with a U.S. address on the account) that will be stored privately by Barnes & Noble. Anytime you make a purchase, the amount will be charged to your card. Keep in mind that you can always change your default card or even input another card for a particular purchase—it's just much easier to have a credit card on file. Choose a valid credit card, type in the account number, and, once it is validated with a quick online check, you're ready to begin using your NOOK Color.

Navigating the Touchscreen

Your NOOK Color can understand six different types of gestures on its screen. If you move your finger up or down, NOOK Color will scroll through its menus. If you swipe your finger left or right, the pages will turn forward or back. If you pinch your fingers in or spread them out, the screen will zoom in or out.

You can also tap the screen to access a book, or to open up the book's menu. The menu will let you change the size of the font, find a specific word, or access the table of contents. You can also quickly tap the screen twice to receive more information about the book at the NOOK Store.

The last feature on your NOOK Color touchscreen is the Press and Hold feature. Press and Holding a word while reading will let you highlight phrases, add annotations, or look up the word in the built-in dictionary. Also, Press and Holding on the front page or within apps will pull up a menu that changes based on what you're currently doing. We'll get into that further in the later chapters on book reading, periodical reading, and using apps.

The NOOK Button

The main menu is accessible by tapping the small "n" or NOOK button. These main menu icons are the most convenient and primary way to navigate through the most used areas of NOOK Color. After tapping the "n" button, the following icons will appear on the bottom of the screen.

- **Home:** This will take you to the Home Screen.
- **Library:** Quick access to the books, magazines, newspapers, comics, graphic novels, and apps on your NOOK Color and any files you transferred to it from your personal computer.
- **Shop:** Takes you to the NOOK Store, where you can make online purchases.

NOOK Color Quick Nav Bar

- **Search:** Lets you look through both the contents of your NOOK Color and the NOOK Store for anything you type in.
- **Apps:** Shows you all the apps installed on your NOOK Color.
- **Web:** Opens a Web browser for surfing.
- **Settings:** Accesses the Settings page to change several NOOK Color features.

The Status Bar

Like most devices, your NOOK Color includes an active area that gives you helpful info like how much battery life is remaining, what you were last reading, and so on. There's no technical name for this area (which you should be aware of if you're speaking with B&N technical support), so in this guide we'll refer to it as the Status Bar.

Located at the top, the Status Bar shows you the following:

- The book, magazine, or item you were just using. Tap the name and it will open to exactly where you left off.
- Your history: Tap to get a listing of the last few items you were enjoying. You can select one of these and jump back in right where you were.

At the bottom of NOOK Color's screen, you'll see the following:

- The placeholder: A marker you left at a particular place in a particular book, magazine, or newspaper. Tap it and it will open up the reading material to where you marked it.
- Wi-Fi, Power, battery status and clock.

Let's take a closer look at the Status Bar.

NOOK Color Status Bar

The Left Side of the Status Bar

The left side of the Status Bar is contextual: That is, it changes based on what you are doing. If you are reading a book, the status information will be much different from, say, if you were listening to music or surfing the Web.

There are several icons that can appear in the bar.

- Open Book
- Download Arrow
- Email Envelope
- Pandora "P"
- "n" update symbol
- NOOK Friends
- Red Notifications circle
- Musical Note

The Open Book icon is a standard bookmark. If you open up your favorite book, magazine, or newspaper, and then go to another screen—such as the home screen—then your NOOK Color will keep your place. You can then tap the Open Book icon and your NOOK Color will take you right back to what you were reading. We'll talk more about reading in Chapter 17, *Books and Periodicals*.

The Download arrow means that your NOOK Color is currently getting a new book, magazine, newspaper, or app for you. With some devices, when you purchase an item you aren't able to do other things while you wait for the download. NOOK Color is great at multitasking, so you can start downloading something and, while it downloads, go back to your reading, Web browsing, or what have you. The Download arrow confirms that your download is still happening. It will disappear when your download is complete. Don't worry if you're not sure how to download things yet, as we'll provide detailed instructions in the following chapters on *Books and Periodicals, Apps, Games,* and so on.

The Email Envelope means that you've got mail. NOOK Color connects to your email, so you can read, reply, and delete when you're on the go. The one thing NOOK Color doesn't do is actually "host" your mail—you'd need a regular email service, such as Gmail, Hotmail, Yahoo!, or another service to run your email. NOOK Color just allows you to access it. That said, the Email Envelope will appear whenever you get new mail. You can sync up NOOK Color to check for new email, so you can stay connected, even while you're away from your home computer or laptop. More info on email is in Chapter 24, *Getting Social.*

The "P" stands for Pandora, the free music service that works well on NOOK Color. Available through the NOOK Store, Pandora streams music—that is, through the wireless Internet—into your NOOK Color. And, because NOOK Color is great at multitasking, Pandora can pipe music through the device while you're reading, Web browsing, or shopping for apps. You know it's playing when the "P" is in the corner. More info on music can be found in Chapter 20, *Music and Video.*

The "n" symbol popping up in the corner may look familiar, as it is the NOOK logo. It means that new software has already been downloaded to and installed on your NOOK Color. While the Download arrow means that NOOK Color is getting one of your requested books, periodicals, or apps, the "n" symbol means that your NOOK Color has downloaded new software from Barnes & Noble. If you're familiar with home computers, you know that hardware companies regularly update their software to eliminate bugs, make improvements, etc. NOOK

Color uses the same system and, as on other devices, NOOK Color software updates are free.

The white icon with two people shows that activity is happening in NOOK Friends. Similar to Facebook, NOOK Friends allows you to follow other people and see what they are currently reading, what apps they are using, and what their opinion is on the latest new reads. In turn, of course, you can share your current opinion and activities, too. A nice perk is that you can also find out what books your friends have available to borrow, plus, you can lend books to your friends. When the NOOK Friends icon appears, there is new activity happening among your friends—it could be a new person sending you a friend request, a message sent to you by a current friend, or another event pertaining to you and your friends. We'll discuss NOOK Friends more in Chapter 24, *Getting Social*.

The red Notifications circle means that you have something new that needs your attention. It could be a new message from a friend or a new book available through LendMe. The number inside the circle tells you how many notifications you have. Tap the circle to see all the notes.

Finally, the red Musical Note icon means that your tunes are playing. In addition to displaying books, your NOOK Color can play your favorite music. You can transfer MP3s from your home computer, and listen to them through your NOOK Color. As we discussed above, your NOOK Color has a headphone jack at the top for standard earphones as well as a set of hidden speakers on its backside for public listening. When you are playing music, the red Musical Note will appear in the Status Bar. Like other activities, you can start up the music player and move on to reading, Web surfing, etc., while the music continues. We'll tell you how to get music on your NOOK Color in Chapter 20, *Music and Video*.

The Right Side of the Status Bar

The first icon, which looks like a fan, is your Wi-Fi icon. When Wi-Fi is active, you'll see little waves coming from the bottom to the top of the icon; if you have any other Wi-Fi-enabled devices, you are probably familiar with this symbol. If the fan is empty, however, then you don't have Wi-Fi right now. Remember that it doesn't mean that Wi-Fi isn't enabled, but that you aren't getting Wi-Fi right now—your local hotspot could be down, or there could be some other technical difficulty. If you're having trouble, try going through the Wi-Fi start-up process again, or skip ahead to Chapter 27, *Troubleshooting*, to get yourself squared away before proceeding.

The second icon, which looks like a little battery, is your Power icon. When NOOK Color is fully charged, the battery will look totally full. As the device is used, the battery will drain. No worries: Your NOOK Color can handle up to eight hours of continuous use and can sit on standby for days. However, when your power finally does get low, the battery meter will shrink and, eventually, flash to warn you that you need to recharge.

The third and final icon, clock, gives you the current time. Tap on it, though, and you can get a list of quick settings:

- Today's date
- Battery life
- Wi-Fi switch
- Mute
- Auto-rotate screen
- Brightness

The battery life gives the actual percentage of remaining power. The Wi-Fi switch, labeled On and Off, allows you to turn off the wireless

NOOK Color Quick Settings

connection. Why would you want to do it? Having the Wi-Fi enabled actually takes more energy, so if you're low on battery power and don't need to download anything, you can turn it off. Also, if you're having challenges getting online, you can try turning the Wi-Fi off for a few minutes and then turning it back on. Again, check out Chapter 27, *Troubleshooting*, for more help with Wi-Fi challenges.

Mute turns off the sound completely on your NOOK Color. You can also tap the Up and Down volume controls to adjust it.

The Auto-rotate screen allows you to customize your view to suit the shape of what you read. Most books, magazines, and newspapers are meant to be read vertically, which is why your NOOK Color is taller than it is wide. However, many can be read horizontally, too. Reading with the vertical view means one page appearing at a time, but reading with the horizontal view means two pages appear across the screen (unless you modify it to read as one very broad column). The horizontal view is great for picture books and other publications that have multipage illustrations. We'll talk more about that in Chapter 17, *Books and Periodicals*. For now, know that NOOK Color will adjust to either a portrait or landscape setting depending on how you hold it. Toggle the Auto-rotate switch to keep the view fixed no matter how you hold the device. A lock icon will also appear momentarily when you open a book so you can decide on your orientation right away.

NOOK Kids books are always shown in landscape format, no matter how you have your orientation set.

Finally, adjusting screen brightness. Tap the Brightness icon and a sliding bar will appear onscreen. Use your finger to increase or decrease the light. The brighter the light, the easier it may be for you to see, but it also takes more battery power. Adjust it as you see fit, balancing your battery needs and your comfort level.

Advanced Settings

Notice the little cog in the upper right-hand corner of the Quick Settings menu? That contains the device and app settings, and a more complete list of controls. Here are the device settings:

- Device Info
- Wireless
- Screen
- Sounds
- Time
- Security
- Keyboard

The app settings are right below the device settings:

- Home
- Shop
- Social
- Reader
- Search

The Advanced Settings page is one of the most powerful pages; here you can tweak and adjust your NOOK Color to truly make it your own. The Advanced Settings page ends up touching on nearly every aspect of NOOK Color from your reading experience to your ability to connect with other NOOK readers, so we'll be referring to this screen throughout the book. At this point, just remember that you can access it either through the cog on the Quick Settings menu or on the Main Menu under Settings.

Warranty Options

Finally, before you start having fun with the device, you want to consider the warranty options. Barnes & Noble has a couple of different ones to choose from:

NOOK Color Protection Plan		
(based on http://www.nook.com/warranty)		
	Standard Warranty (1 year)	**B&N Protection Plan** (2 years)
Customer Service	x	x
Rapid Replacement	x	x
Accidental Damage	x	
Extended Service	x	

The best part is that you automatically got the basic protection as soon as you bought your NOOK Color! It's just a matter of deciding if you'd like to upgrade to the advanced protection plan.

Both plans include:

- Free customer support
- Rapid replacement if NOOK Color malfunctions
- Minimum of one year protection

However, the B&N Protection Plan adds:

- Two years of protection
- Rapid replacement for accidents like spills and cracks

Visit http://www.nook.com or call 1-800-843-2665 for the latest B&N Protection Plan pricing. For more details on the support plans, read the FAQ at the end of this book.

Now you have everything you need to get started with your NOOK Color. But what about reading books, downloading items, and managing your ever-growing collection? The rest of this section of the book is dedicated to making sure you get the most out of your device. Let's get started.

16. NOOK Store

While there are a few books—such as Bram Stoker's *Dracula*—included on NOOK Color from the get-go, you probably want to read other books and expand your digital library. This is where the NOOK Store comes into play.

The NOOK Store is where you will be buying, sampling, and browsing through the vast array of books available for reading on a NOOK Color. The shop features all of the books and categories you'd be able to find at a Barnes & Noble retail store. After all, what good is an eReader with nothing good to read? To access the NOOK Store, push the NOOK button and touch the Shop icon.

The NOOK Store's Front Window

There are two main sections on the NOOK Store's front window: Book Genres and Popular Lists. If you're looking for a specific type of book, scroll through the Book Genres.

The Popular Lists pane covers similar ground. This section reveals greater detail about the current major titles in the literary world. Here, you'll find things like the *New York Times* bestsellers and new releases. It also lists featured books on sale.

Browsing

Browsing through the Store is easy. By using lists, or simply by searching, countless books are waiting to be discovered. If you aren't sure what books to look for, it's best to start with a list, or simply search for a specific book you want. Just tap the list/section you want and start narrowing your search. If you don't have a specific titles in mind, we'll tell you how to find a book through the first pane.

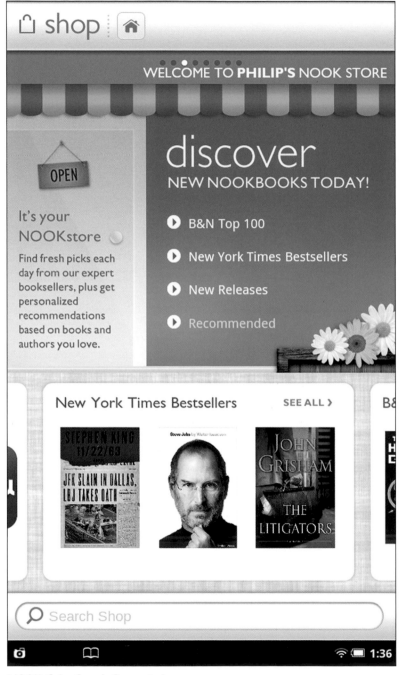

NOOK Color Store's front window

Books

If you already know what book you're looking for, then use the Search Bar at the bottom of the screen. Type in either the name of the book or the author, and you'll find a list of works containing the words you've typed. This function works like any other search engine. If you're browsing for something new and exciting to read, however, we recommend you go through the Store's detailed list to find something that piques your interest.

Tap the Books icon on the Store's home screen and you'll see a page with a scrollable collection of lists at the top and several scrollable shelves of recommendations at the bottom. Each category covers a different literary genre. From biographies to humor, every style is represented.

Scan each page until you see a genre that looks interesting and select it. The next section will contain another series of pages detailing a more specific kind of book. For example, selecting Humor will lead to a list displaying the variety of humor books, like essays or cartoons. Here, you'll find a countless supply of books directly related to your search. Browse until you find a book that looks interesting. Tap on a book's cover and a brief summary of the book and publication info will appear. If you want to buy it, tap the button with the price on it. Then, the button will change to say Confirm. Press Confirm, and the book will automatically be downloaded to your NOOK Color. If you aren't sure about buying the book, look to see if the publisher has made a free sample available, and if so, download it by pushing the button marked Sample. After you've decided whether or not to purchase the book or download the sample, simply press the Back button—the little button with an arrow pointing left—and look for other books to check out.

Learning how to buy books is obviously central to enjoying your NOOK Color. After you've learned how to find books, try downloading a few samples. It won't take long before you have a full library of classics and new bestsellers.

In our next chapter, we'll talk about how to access and read all the books you've just sampled and purchased.

Periodicals

You can browse and shop for magazines and newspapers similarly to books. The difference, however, lies in purchasing them. Instead of being able to download a sample, with all publications you can try a free 14-day trial, or simply purchase the current issue. Once you are certain you want a particular magazine or newspaper, you can also subscribe to it for a subscription fee, and each new issue will automatically download to your NOOK Color as soon as it is released. There is usually a steep discount for subscribing compared to buying each individual issue, and then there's no stress about ever forgetting to download one. It guarantees a continuous supply of new content for your NOOK Color.

Apps

NOOK Color users can also download apps for their devices. These are programs that work like tools that tell the weather, or even games like Scrabble or crossword puzzles. Navigate through the Apps Shop until you find one that sounds compelling, push the button with the price on it, then Confirm, and the app should be available to use within seconds. In order to know more about the variety of available apps, we've included an entire chapter about them later in this section.

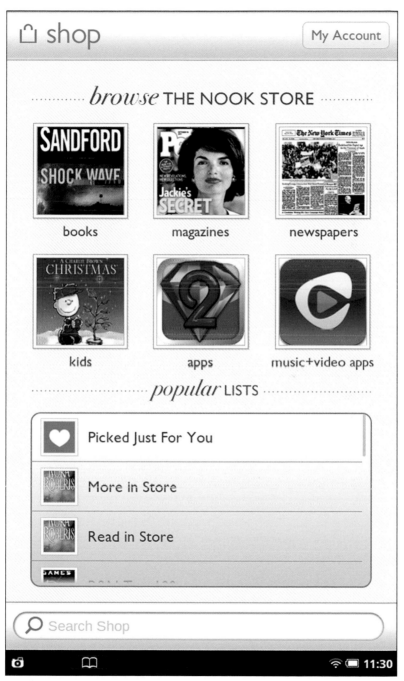

NOOK Store's front page

17. Books and Periodicals

Now that you've purchased some exciting new titles and are on your way to building an impressive library on your NOOK Color, this chapter will show you how to dive into these great new reads and enjoy special reading features like bookmarking, looking words, and more. We'll also discuss how to navigate and organize your Library by creating your own personal bookshelves and making books, magazines, and newspapers quick and easy to find.

Finding Books and Navigating the Library

Your NOOK Color Library can be accessed by pressing the NOOK button to open the Quick Nav Bar and then tapping the Library icon. When you first tap the Library icon, a list will appear with all the reading you have downloaded to your NOOK Color. This list can be managed with the two navigation menus, found immediately under the Library title. The first menu helps navigate through books, periodicals, apps, and any files you have transferred from your personal computer; we'll explain this further in Chapter 25, *Memory and Storage*. The second menu reorganizes the list by title, author or the most recently added books in the list.

If you don't like the default layout of the Library, you can change it to look like a simple list of books. Near the top right corner of your NOOK Color's Library screen, there are two buttons. One looks like nine squares arranged like a grid and the other looks like a list. Pushing either of them changes the appearance of the Library, so choose whichever one suits your preference.

To locate a book, simply scroll through the page until you find the book you want to read. To scroll, swipe your finger along the screen. Tap the book's cover and the book will open. You can also open magazines, newspapers, and PDF files by tapping on their covers.

If you're looking for a book you know you have, but it isn't in your Library, be sure to check out your Archives. Archived books are books that are not on your NOOK Color but Barnes & Noble recognizes that you own them. To get to the Archive, tap the "my stuff" icon in the Media Bar and then tap "Archived" in the drop-down menu. Tap the button marked Archived and tap the "Unarchive" button to re-download items. To learn more about this, check out the Chapter 10, *Memory and Storage*.

Reading Books

Once you have opened a book, diving in to start reading is as easy as a tap or a swipe. Let's start with turning pages. The first way to turn a page is to tap along the right or left margin of a page, along the very right side or left side of the touchscreen. You can also turn a page by swiping your finger from right to left, or left to right, to turn the page back.

Once you master page turning, you'll be able to navigate through anything NOOK Color has to offer. There are, however, many other features inside a book that are worth discussing. NOOK Color lets readers annotate their books, and features several functions to expedite page-turning searches for specific parts of the book. You'll find more about other ways to personalize your NOOK Color in the next section.

Notes and Highlight

Sometimes, something you read is so striking, you'll want to remember exactly where you read it, and that might not seem like the easiest thing to do on an eReader. After all, you can't fold down a page corner on NOOK Color. Fortunately, NOOK Color has several key features that far surpass simple page-folding.

Notes and Highlight are very similar, except the former lets you add a few quick comments. Essentially, both features let you mark a word or passage, allowing you to access it from the Content page. This makes looking for specific parts of a book extraordinarily easy.

By holding your finger down for a second on the text in a NOOK Book, a horizontal menu pops up (the Text Selection toolbar) that lets you do several things to that selected text. You can highlight the text, add an annotation, share it with friends, post it on Facebook or Twitter, or look a particular word up in NOOK Color's built-in dictionary.

Highlighting a word or passage helps mark key points of the book that you may want to return to later. Adding notes to the highlights also lets you add any comments you'd like to remember while reading the book. If there is a note on a page, it will be marked with a logo resembling a little note on the right side of the page. Tapping it will display the note you left.

Bookmarks

Bookmarks, as in printed books, mark an entire page for future reading. Bookmarks don't single out noteworthy words or allow annotations. Instead, they simply mark a page so you can easily return to it later.

When you tap a page in a NOOK Book, a small tab that looks like a ribbon with a lower case "n" appears in the top right corner. Tapping it highlights the Bookmark, and this page number will be saved for future reading. To undo a Bookmark, simply tap it a second time, and ribbon should disappear.

you are going to?' She was in such evident distress that I tried to comfort her, but without effect. Finally she went down on her knees and implored me not to go; at least to wait a day or two before starting. It was all very ridiculous, but I did not feel comfortable. However, there was business to be done I could not allow anything to interfere with it. I therefore tried to raise her up, and said, as gravely as I could, that I thanked her, but my duty was imperative, and that I must go. She then rose and dried her eyes, and taking a crucifix from her neck offered it to me. I did not know what to do, for, as an English Churchman,[i] I have been taught to regard such things as in some measure idolatrous, and yet it seemed so ungracious to refuse an old lady meaning so well and in such a state of mind. She saw, I suppose, the doubt in my face, for she put the rosary round my neck, and said, 'For your mother's sake,' and went out of the room. I am writing up this part of the diary whilst I am waiting for the coach, which is, of course, late; and the crucifix is still round my neck. Whether it is the old lady's fear, or the many ghostly traditions of this place, or the crucifix itself, I do not know, but I am not feeling nearly as easy in my mind as usual. If

39 of 433

 11:25

Selecting text in a NOOK Book

To access the page later on, go to the Content menu. The Bookmarks tab is the one closest to the right. Tap it, and a list of Bookmarks will appear on your screen.

Navigating NOOK Books

NOOK Books are extraordinarily easy to navigate. In addition to the above-mentioned Notes and Highlight, NOOK Color has several features to make finding the exact part of a book a simple as a tap.

In order to start, tap on the center of any page of the book, and you'll find the following reading tools:

- Slide bar
- Content
- Find
- Share
- Text
- Brightness
- Discover

Even with all of the in-book options, you can always tap the Quick Nav button and navigate NOOK Color like you normally would. This will take you out of the book and into other pages like Settings or the NOOK Store.

Slide Bar

Notice the small slide bar at the bottom of the NOOK screen? This bar lets you navigate quickly to a specific page by simply moving your finger left or right along the bar; the book will either advance forward or back several pages, depending on how far you moved your finger. Once you let go, NOOK will tell you some brief details about the page, and usually how many pages are left in that chapter.

Dracula (Barnes & Noble Classics Series)

you are going to?' She was in such evident distress that I tried to comfort her, but without effect. Finally she went down on her knees and implored me not to go; at least to wait a day or two before starting. It was all very ridiculous, but I did not feel comfortable. However, there was business to be done, and I could allow nothing to interfere with it. I therefore tried to raise her up, and said, as gravely as I could, that I thanked her, but my duty was imperative, and that I must go. She then rose and dried her eyes, and taking a crucifix from her neck offered it to me. I did not know what to do, for, as an English Churchman,[i] I have been taught to regard such things as in some measure idolatrous, and yet it seemed so ungracious to refuse an old lady meaning so well and in such a state of mind. She saw, I suppose, the doubt in my face, for she put the rosary round my neck, and said, 'For your mother's sake,' and went out of the room. I am writing up this part of the diary whilst I am waiting for the coach, which is, of course, late; and the crucifix is still round my neck. Whether it is the old la-

CHAPTER I 10 pages left in this chapter

Go Back

39 of 433

Go to Page

content find share text brightness discover

11:25

The Reading Tools menu

Content

The Content button brings you back to the table of contents. Here, you can easily jump to each chapter in the book and navigate through the list in the first pane to see what each chapter is called and on what page the chapter begins. Tapping on the chapter name takes you to that specific chapter.

The Content screen also has tabs in the upper section of the page that you can use to navigate through your Content, Notes, Highlights, and Bookmarks. To look through them, find the other tabs near the top of your NOOK Color screen and tap them. The lists will look similar to the Chapter pane, but will feature all the Notes, Highlights, and Bookmarks you've already made in the book. These are great ways to personalize the navigation experience.

Find

The Find feature lets you look for any word in the book. Just type in the word and within seconds, you'll see a list of places where the word appears in the book. The list displays part of the sentence in which the word appears, as well as the page number.

To use the Find feature, open up the Reading Tools menu and tap the button marked Find. This will bring up a Find Bar. Type in the word you want to look for, tap the "Search" button in the lower right corner of the keyboard, and a menu similar to the ones found in the Content feature will appear, listing every appearance of the word you just typed. From here, tap the section of the book you want to go to and NOOK Color will take you there.

These features don't only work for NOOK Books. The Reading Tools menu works exactly the same way in PDF files (though PDFs saved as images aren't searchable) and most periodicals.

Text

There are five different options in the Text Settings menu. The first one—located in the middle of the screen—changes the font size. This changes the size of the words on the pages. The next tab changes the style of the font. This affects the visual appearance and flair of the words. Fonts include Trebuchet and Gill Sans. The following option allows you to change the color of the text itself. Next, the paragraph icons change the line spacing and the margins around each page. This changes how closely spaced the words are to each other and how close they are to the edge of the screen.

Adding books from places outside of the Barnes & Noble Bookstore is a great way to add books to your NOOK Color. There are many free books available for your NOOK Color, including NOOK Books of works in the public domain (works that are no longer copyrighted) and other free eBooks available from places other than Barnes & Noble. (Note, however, that all eBooks must be in EPUB or PDF format to be read on a NOOK.) We'll talk about getting books from sources other than the NOOK Store later in this book, and address how to add them to NOOK Color in Chapter 25, *Memory and Storage*. We'll also talk about where to download free books in the appendix.

The final option, the toggle switch on the bottom right, lets you set all of these options to the publisher's defaults (if provided). In most cases, the publisher's defaults set everything the way the book-publishing company wants it, setting the font, margins, and line spacing the way the publisher intended.

Creating Bookshelves

To best organize and manage your NOOK Library, you'll probably want to create your own bookshelves. This way, you can personally manage the books any way you want. To do so, start by going to the Media Bar in the Library, tap the My Stuff button and in the pull-down menu tap on My Shelves.

From here, tap the Create New Shelf button in the upper left. This opens a page that will let you name the new Shelf. We recommend that you use a name that will be easy for you to remember and understand later, such as a Shelf named for each literary genre you like or the month you purchased the book. Type in the Shelf's name, and tap Save in the bottom right corner.

After you name the new Shelf, it will appear at the bottom of the screen—if you don't see it right away, scroll down the screen with your finger. Tap the Edit icon and a list appears of all the content currently on your NOOK Color. Go through this list and check each book you want to place on that Shelf by tapping the check box next to the book's title. A check mark should appear in the box if you tapped it correctly. Once you've checked each book you want on the Shelf, tap Save, and the new Shelf will appear with your added books on it.

If you ever want to add anything to the Shelf, tap the Edit button right under the Add Shelf button. The Shelf screen only shows

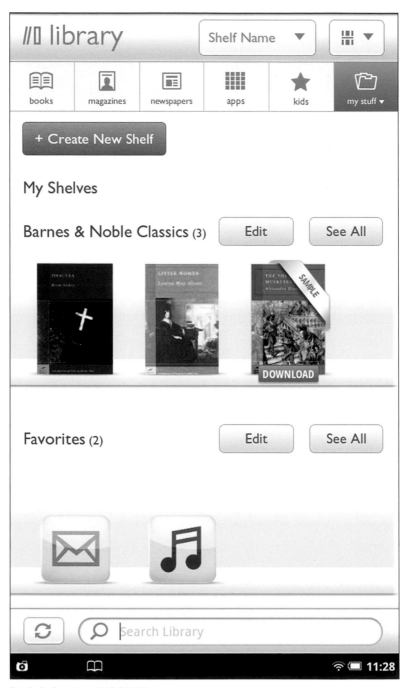

Bookshelves in a NOOK Library

four books on its front page, but you can tap the See All button to see all the books on the shelf.

Borrowing Books

One of the biggest complaints about digital books is that, unlike traditional books, you can't borrow them. And one of the biggest benefits of using NOOK Color is that you can borrow books!

There are two ways to borrow books:

- From a NOOK Friend
- From your public library

With NOOK Friends, you can borrow books from your connected friends and lend books to them, too. All it requires is sending an invitation to a real-life friend via NOOK Color. To get connected, your friend needs to:

- Be a registered user at the Barnes & Noble Web site
- Have a registered NOOK device or NOOK eReader app including NOOK for iPad, NOOK for Android, NOOK for iPhone, NOOK for PC, and NOOK for Mac.

To learn more about NOOK Friends, check out Chapter 24, *Getting Social*.

Periodicals

One of the best features of NOOK Color is its ability to download periodicals, giving you easy access to the latest newspapers and magazine issues. A new periodical you've subscribed to will download automatically when you're connected to Wi-Fi. The NOOK Store offers plenty of magazines and newspapers from cities big and small.

To purchase a periodical, navigate through the Store's menus—or simply search for a specific magazine or newspaper—and tap either Subscribe or Buy Current Issue. Subscriptions typically cost significantly less than the price of buying individual issues month after month. As discussed in the previous chapter, if you want to sample a magazine or newspaper before you subscribe, many periodicals offer a free 14-day trial subscription that will let you test-drive the publication before committing to a full-year subscription.

Reading Periodicals

By now, you're probably used to reading books on NOOK Color and magazines aren't too different. After purchasing a magazine or newspaper, go to the Library and tap the one you want to read. The magazine's cover should be the first thing to appear.

Periodicals on NOOK Color are designed for ease of navigation. The biggest difference between periodicals and books is that newspapers have subsections, each of which has its own table of contents. After purchasing an issue of a newspaper, look through the table of contents for a section that seems interesting. Search through the pages of stories until you find one that you would like to read, tap the headline, and you'll be presented with the article. Once you've finished reading the article (it may be several pages long) the bottom of the last page will show Next Article and Previous Article buttons (which also appear on the first page of articles). Tap them to access the next or previous pieces. Or, you can return to the table of contents by tapping anywhere on the screen to access the Reading Tools menu, and then tap Content.

Periodicals are subscription-based. In other words, your credit card will be charged on a monthly (or bi-monthly, depending on the publication) basis for new issues.

Archiving and Deleting Periodicals

Periodicals are updated regularly, so it's easy to be overwhelmed by the amount that can clutter up a NOOK Color's library. In order to avoid this, you can archive periodicals the exact same way you can archive a book. In your Library, if you double tap the periodical you want archived, you'll be brought to the Detail page, where after tapping "Manage," you will be given two options:

▶ Archive
▶ Delete

Archive will do just that—the magazine or newspaper will be removed from NOOK Color and placed in the B&N Archive until you want to re-download it, which we'll cover in "Managing Space," in Chapter 25, *Memory and Storage*.

Delete will remove the periodical completely from your device and from your storage. There will be no trace of it on your account aside from the receipt of purchase.

Unsubscribe from a Periodical

Unfortunately, there isn't yet a way to unsubscribe from a periodical directly from NOOK Color. The only way to do so is by going through Barnes & Noble's Web site. From the Web site, however, it is a very simple process to unsubscribe.

Simply go to BN.com and log in with your account, click on Manage Subscriptions, and click the Cancel Subscriptions button for each periodical you want to unsubscribe from. You'll be able to keep the issues that were downloaded during the time you subscribed, including those downloaded during the free trial.

Reading a periodical

18. NOOK Kids

One of the biggest benefits to reading with NOOK Color is the rich, colorful images that come to life on the screen. Kids books are awesome on NOOK Color, allowing you to enjoy new and even classic titles in a whole new way.

The first step is downloading a NOOK Kids book. You can open up one of the NOOK Kids books that have been preloaded onto your NOOK Color or, if you like, tap the "n" key, use the Quick Nav Bar to go to the Store, and search for a favorite title.

Once your NOOK Kids book has successfully downloaded, tap the cover to open it. Notice how it goes horizontal? Most NOOK Kids books (excepting some chapter books) are presented in landscape mode. Books for little kids usually have big, bright pictures, and the wider view allows better detail than the vertical setup. Turn your device so it sits in both of your hands.

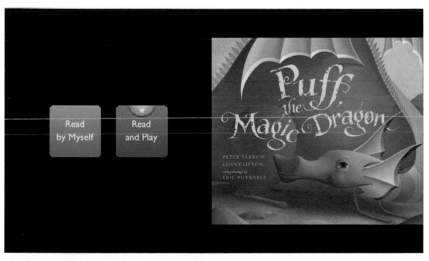

NOOK Kids books can be read or read aloud

To the right is the full-color book cover. To the left there are three possible reading options:

- ▶ Read by Myself
- ▶ Read to Me
- ▶ Read and Play

Let's take a closer look at these three very different ways to read

Read by Myself

Tap the Read by Myself option. You can now use your finger to slide from page to page: Push to the right to go forward and push left to go back to the previous pages. Again, all children's books show a two-page spread, so you can quickly go from one two-page section to the next. The two-page format makes perfect sense when you realize most NOOK Kids books have a page of text facing a page of illustrations or photos. Some books, like those from Dr. Seuss, have a combination of text and illustrations across both pages!

Reading a NOOK Kids book

There is a little arrow at the bottom of the screen, nestled right where the two book pages meet. Tap this arrow and thumbnail images of all the book pages will appear in the bottom half of the screen. If you like, you can skim through the book quickly by sliding the thumbnails left or right with your finger. It is a great way to skip to your favorite part of the book.

Read to Me

Many NOOK Kids books also have the Read to Me option. Go back to the book cover, either by sliding right until you get to the front of the book or by tapping the white arrow and sliding the thumbnails to the cover. Once you get there, tap the Read to Me option.

Now the book will be read to you by your NOOK Color! If you need to adjust the sound, turn your NOOK Color upright and use the volume Up and Down buttons on the right side of your device to make it higher or lower.

> There is no standard narrator voice for NOOK Kids books, so each has its own unique sound. Try different books to see what the narrator will sound like.

The narrator will read what's on the two pages slowly and clearly. It will then wait until you flip the page. Want it to read the previous pages again? Just flip to those pages and the narrator will read them for you. You're in control of what you want it to read, so, unlike audiobooks, there's no reason to feel rushed.

Not every NOOK Kids book has the Read to Me option, but you'll be able to tell if it is available when you get to the cover page.

Read and Play

The final level of book interactivity is the Read and Play option. Again, if you're in the middle of a book, hop back to the cover page by sliding the pages or using the thumbnails. Tap on the Read and Play option.

At first Read and Play books seem like standard titles, but they have special spreads, marked by a star, that allow for cool interactivity. For instance, the book may ask you to tap the screen to touch a particular character or pinch an area of the screen to do a fun action. In these special sections, the pages will not progress with the traditional slide motions—the book will suspend page flipping so you can concentrate on the featured page spread.

Zooming in

Sometimes you want to see bigger text or get a closer look at a detailed picture. You can use your fingers to pinch the screen and zoom in or out. Touch the screen with your thumb and index finger, and spread them wide to zoom in. Do the reverse—push them closer—to zoom out.

Keep in mind that the zoom-in function doesn't work on the special Read and Play segments.

19. Web Browser

NOOK Color is great for reading books, magazines, and other periodicals. It can do much more than that, though. The rest of the NOOK Color chapters in this book will focus on its more advanced capabilities.

Surfing the Web

One of the most exciting features of NOOK Color is its full-fledged Web browser. NOOK Color can surf the Web as easily as your average computer.

Here are some details to keep in mind:

- You can only use the browser while connected over Wi-Fi.
- The keyboard pops up on the touchscreen.
- Unlike many other tablets, NOOK Color can play Flash video and animation.

To open up the browser, tap the NOOK button to open the Quick Nav Bar. The screen menu will pop up with the following options:

- Home
- Library
- Shop
- Search
- Apps
- Web
- Settings

Touch the Web icon and you're ready to surf. Feel free to hop online and play around with the Internet. Next, we'll take a closer look at the browser and show you how to get the most out of Web surfing on NOOK Color.

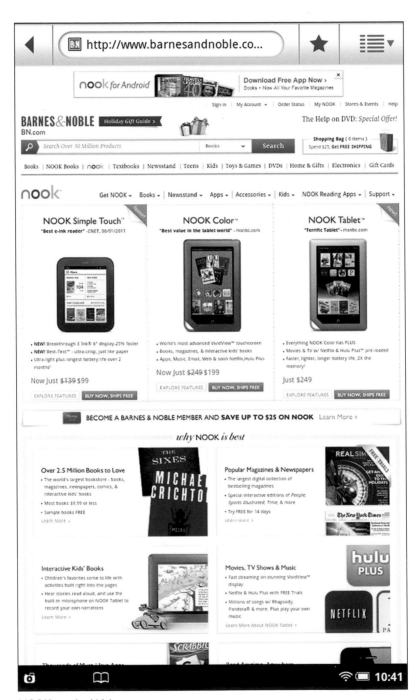

NOOK on the Web

Browser Overview

If you're familiar with the World Wide Web, your NOOK Color Internet browser will be easy to use. In fact, it offers the same experience you'd have on your home computer or laptop. The only adjustments are the touchscreen keyboard and the screen (which can show the browser in both vertical and horizontal orientations). We'll discuss those differences in a second.

For now, let's take a tour of your Web browser. From the top, you'll see the following icons:

- Back Arrow
- Web Address Bar
- Bookmark Star
- List of Options

The Back arrow will open up the previous Web page. Don't worry about tapping it now, as you haven't been to any other Web pages, so there are no previous Web pages to open! If you tap it when there are no older Web pages to return to, NOOK Color will close the browser and send you back to your NOOK Color Home Screen.

The Web Address Bar shows the current Web page. You can type in your favorite Web address here. The Bookmark Star is a quick link to all your marked places. It also stores your most visited places and your history.

List of Options reveals all the details you can tweak to personalize your Web-browsing experience. The full list of options will be discussed later in this chapter.

NOOK Color's Web browser

Your NOOK Color Web browser automatically opens up to the official Barnes & Noble NOOK page. Here you'll find lots of useful stuff, including interactive tutorials, descriptions of the latest NOOK Color accessories (which we'll touch on at the end of this section), and NOOK news via the NOOK blog.

There are three primary ways to browse a page:

▶ Tap
▶ Double tap
▶ Slide

Tap on an item, like a Web link or a video, and NOOK Color will open the Web page or play the video for you. Double tap your finger and NOOK Color will zoom in on a particular part of the page, or, if it is already zoomed in, will zoom out to show a bigger area.

Slide your finger up and down the touchscreen and the screen will scroll accordingly. Try touching the different links and exploring this Barnes & Noble page to get a feel for the browser. Next, we're going to visit your favorite Web pages. Like your computer's browser, there are two ways to get to a Web page:

▶ Do a Web search.
▶ Type in the Web address.

Tap the Web Address Bar and you'll see the keyboard pop up. Type in a topic or Web page name and your NOOK Color will use Google search to find it online. For instance, type in "Emeril Lagasse," press Go, and Google will give you a listing of everything related to the famous chef. It is literally the same information you'd get on a traditional computer, so you can feel comfortable that it is accurate and up to date. Also, if your topic is popular enough, Google will offer suggestions on what you are looking for as you're actually typing in the topic.

You can also type in a specific Web address. Touch the Web Address Bar at the very top of the screen and NOOK Color will allow you to input the location. It is also smart enough to know that a "http://www" belongs at the beginning of most Web addresses, so you won't need to type that part in most of the time. For example, type in "nytimes.com," and NOOK Color will display the home page of the popular *New York Times* Web site.

Browser Details

Now we'll look at two ways NOOK Color can make your Web experiences better.

Magnified Type

Whatever Web site you are on, slide your finger on the touchscreen. Notice the "+/-" symbols appearing in the lower right-hand corner? Try touching the "+" symbol. The size of the Web type will increase. Touch the "-" symbol and the Web type will get smaller.

Bookmarking

Bookmarking is equally important. Touch the Bookmark Star located at the top of the screen, right next to your Address Bar. Your NOOK Color will show you three tabs:

▶ Bookmarks

▶ Most Visited

▶ History

The Bookmarks tab shows you all your favorite places on the Web. You don't have any bookmarks yet, but you can add one right now. NOOK Color will suggest different bookmarks, including your current Web page. If you want to add the Web page, look for the icon showing the name of the page and tap the Add icon. NOOK Color will ask you if you have a special name for the Web page and will prompt you to confirm the Web address. Touch OK and the Web page will be bookmarked for you. Tap the View option in the lower right-hand corner to toggle between the word list view and the visual thumbnail view of your bookmarks.

Most Visited shows the places you've browsed most frequently. Tap on one of them to visit the page now. See a page you'd like to add to your bookmarks? Just tap the Bookmark Star next to the Web address and your NOOK Color will add the page to your bookmark list. You can also tap History in the lower right-hand corner to clear your recorded Web browsing history.

The History tab gives a complete listing of every Web site you've visited. Again, tap the Bookmark Star next to any Web site and NOOK Color will bookmark it for you. You'll notice that the list has headers like Today. NOOK Color organizes your history based on day. Tap the Today headline and NOOK Color will hide the history behind the headline so you can easily see other days. As on the Most Visited page, you can tap the History option in the lower right-hand corner and clean out your Web browsing history.

Advanced Options

There are also a lot of advanced options, all hidden underneath the List icon in the upper right-hand corner. Get back to your Web browser, tap the List icon, and you'll see these options:

- New Window
- Bookmarks
- Windows
- Refresh
- Forward
- More Options

Like other computers, your NOOK Color Web browser allows you to have multiple pages open. Tap the New Window option, and it will open a new browser page.

Don't worry: The previous page didn't disappear! NOOK Color just tucked it away so you can explore a new page. Tap the Windows option and you'll get a list of all the windows you have open. Hit the "X" icon next to any of the open windows and NOOK Color will permanently close it. Tap on the name of the Web page to reopen the window.

Refresh will go online and make sure your Web site is updated with the latest info. For instance, if there is breaking news on the *New York Times* Web site, you can tap Refresh to have your NOOK Color update the page with the most current reports.

The More Options section reveals additional selections:

- Add Bookmark
- Find on Page
- Page Info
- Downloads
- Settings

Find on Page allows you to look for a specific term. Tap it and the keyboard will pop up. Type in a term that you think appears on your current Web page, and NOOK Color will find it. Next to the term, your NOOK Color will show the number of matches as well as forward and backward arrows so you can jump between the different places where the term appears within the text. If the term doesn't appear in the text, NOOK Color will say "o matches."

Page Info gives you the name, Web address, and any other public information about the current Web page.

Downloads lists any media you have downloaded onto your NOOK Color. It will give you the status on all of them, including how much longer it will take to download an item.

Finally, Settings sends you to the browser portion of the Settings menu located on your NOOK Color Home Screen. The options here include Text Size, Image Details, and Zoom.

20. Music and Video

The previous chapter discussed Web browsing, one of the many NOOK Color benefits that goes well beyond reading books and magazines. Another huge perk is that your NOOK Color can play some of your favorite multimedia. Multimedia means most of your favorite music and videos.

What you'll need is:

▶ Your NOOK Color power cable

▶ A PC or Mac with a USB port

Transferring Music and Video

Ready to get some songs and videos onto your NOOK Color? Before you do, keep in mind that your NOOK Color can't understand every file you throw on there.

Files That Work

Your NOOK Color can play these types of music files:

▶ AAC

▶ amr

▶ mid

▶ MIDI

▶ MP3

▶ M4a

▶ oog

▶ wav

NOOK Color cannot read WMA (Windows Media) files.

As far as video, your NOOK Color understands these formats:

▶ Adobe Flash

- 3gp
- 3g2
- mkv
- mp4
- m4v
- MPEG-4
- H.263
- H.264

It cannot read:

- MOV/qt
- AVI
- Xvid/DIVX
- WMV/VC-1

You can always use trial-and-error to see if your NOOK Color will read a file, but there's a quicker way. Turn on your computer—PC or Mac, it doesn't matter—and find the file you'd like to enjoy on your NOOK Color. If you have a two-button mouse, highlight the file and click the right mouse button. You will see an option called Get Info on a Mac or Properties on a PC. Select that option and you'll get a list of details about the file. See File Type and you can check if your file is on any of the lists above. If you really need to make a file playable on your NOOK Color, we recommend going online and downloading a file converter so you can turn the current file into one that's compatible with your device. There are dozens of free converters available on the Web.

Getting Your Files on NOOK Color

Once you find a file you can transfer, here's the process:

- Disconnect the power plug from your NOOK Color power cable.

- ▶ Plug the power cable into your NOOK Color.
- ▶ Plug the other cable end into your computer's USB port.
- ▶ Wait for the disk icon "MyNOOKcolor" to appear.
- ▶ Drag and drop the file(s) onto the appropriate folder in My Files.
- ▶ "Trash" the NOOK icon and disconnect NOOK Color from your computer.

Let's go through each step. First, as you remember from Chapter 15, NOOK Color's power cable comes in two pieces: the plug and the cable. Now you'll want to disconnect them again. Grip the power cable right where it meets the power plug and give it a good tug. Put the power plug to the side for now.

Second, plug the opposite end of the cable into your NOOK Color. Nothing fancy here, as this is the same thing you do when you plug in the cable to charge up your device.

Third, connect the remaining end (where the power plug was just connected) into your computer's USB port. As you can tell, the USB plug has a thin, wide base. Look for the corresponding hole on your computer. If you have an older PC with a tall tower, you probably have at least one USB port in the front and in the back. On most modern desktop PCs and Macs, the USB ports will be in the back of the computer. And if you're a laptop user, the USB ports will be on either the right or left side of the computer.

Fourth, you'll know immediately if you've found the right hole and plugged it in securely: Your NOOK Color will light up and a drive icon called MyNOOK Color will appear on your desktop. Use your mouse to double click on this NOOK symbol. It will show several folders, but you are only concerned with the My Files folder. Here is where you can drag and drop your music and video files into NOOK Color.

Go deeper into the file and you'll see the following folders:

- Books
- Documents
- Magazines
- Music
- Newspapers
- Pictures
- Videos
- Wallpapers

Fifth, drag and drop your music or video file into the appropriate folder. Remember to ensure that the file type is included on the list of compatible types outlined earlier in this chapter.

Sixth, eject NOOK Color from the computer. It's not a matter of just unplugging it, as the computer needs to be given a heads-up that you're removing the device from it. It depends on your computer, but most PCs and Macs allow you to drag and drop the actual NOOK icon from the desktop to the Trash Can/Recycle Bin icon. This lets the computer know that you are going to unplug the device.

Now your new files are in your Library. To get there, tap your NOOK button, then touch the Library icon. You'll see a listing for several items:

- Books
- Magazines
- Newspapers
- My Shelves
- My Files
- LendMe

Tap on My Files. You'll see a folder called My Files. Touch the folder and the same file folders listed on your computer (Books through Wallpapers) will be listed on the screen. It is a mirror of what you dragged and dropped on your computer.

Tap the square icon in the upper right-hand corner to toggle between the visual thumbnail view and the word-focused list view.

My Files in Library

You can also access your music through the Music Player app.

Listening to Music

You can listen to two types of music on your NOOK Color:

 ▶ Your music

 ▶ Streaming music

Your music consists of the tunes that we've just dragged and dropped into NOOK Color from your computer. However, since your NOOK Color is connected to the Internet, it can also stream music. Streaming music means your NOOK Color will download the songs live while you listen, kind of like an Internet version of the radio. Pandora and other music-streaming apps make it easy (and usually free!) to stream music.

Before we get into Pandora's streaming music, take a look at how you can listen to your own music catalog.

Listening to Your Music

Now that you're in the My Files folder, tap on the Music folder. You'll see your music listed.

Tap on a song. If it is the right file format, it should start playing immediately. The song will play through the built-in speaker, but make sure to adjust the volume before putting on any headphones— you don't want to hurt your ears! The volume can be adjusted with the two Up and Down buttons on the right side of your NOOK Color, just below the headphone jack. If the volume is turned down, you can tell music is playing when a note icon appears in the lower left-hand corner of your screen.

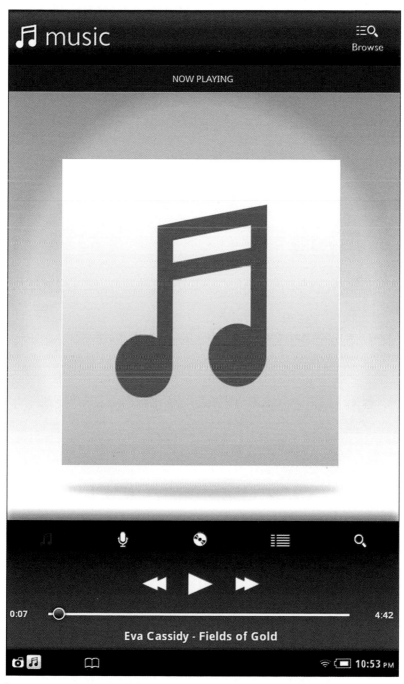

Listening to music on NOOK Color

The majority of the screen is taken up by the song or album cover art. You'll notice several icons immediately below the cover:

- Shuffle
- Repeat
- Album Cover View
- List View
- Browse

Indicated by the weaving arrows, the Shuffle icon toggles on and off. When Shuffle is on, your NOOK Color will mix up the order of the current list of songs.

The circular Repeat icon has three modes: Off, On, and Single. If Repeat is off, the music will stop once the song or album ends. If Repeat is on, the current song or playlist will start over again. If Single Repeat is on, it will repeat only the current song.

Album Cover View and List View will toggle between the album cover art and the text-based list of songs. Tap the magnifying glass and you can browse for a particular song or album in your Library.

Below the first row of icons is a second set of controls:

- Rewind/Restart
- Play/Pause
- Fast Forward/Skip
- Timeline

Like a CD or DVD player, NOOK Color allows you to move forward or back within the current song by using the Rewind, Play, and Fast Forward buttons. Tap the Play button to pause or resume the music. Holding the Rewind or Fast Forward button will skip a few seconds back or ahead in time, while a quick tap of either will restart or skip the current song.

If you want more control, use the Timeline located below these icons. Touch the moving dot. Now you can slide the dot up and down the Timeline to go to a particular part of the current track. The time remaining and the total track time appear next to the Timeline. You'll see the name of the current track below the Timeline.

You can also make playlists using the music on your NOOK Color. Switch over to your List View (the fourth icon right below the album art). Now, find a song you would like in your new playlist and hold your finger down on the name. The following menu will appear:

- ▶ Play
- ▶ Add to Playlist
- ▶ Delete
- ▶ Search

Select Add to Playlist, then select New. Your NOOK Color will ask you to name the playlist. For instance, you might name a playlist Stephen King Reading Mix. Now you can find another song you'd like to add to your playlist, hold down your finger on the name, and add it to the Stephen King Reading Mix.

To access your playlist, tap the Browse icon in the upper right-hand corner, then touch the List View icon. Find your playlist and tap on the name. Now your current playlist will appear. Tap the first song title to start the playlist.

Not happy with the order of the songs in a playlist? Find the song you'd like to move, hold your finger on the triple-line icon to the left of the song title, and drag it to where you'd like it to be in the playlist.

Streaming Music with Pandora

As we mentioned earlier, your NOOK Color can access music through the Internet, too. Here's what's required:

▌ A reliable Internet connection

▌ A music-streaming app, such as Pandora

> If you're not sure how to download an app to your NOOK Color, flip to Chapter 21, *Apps*.

Why do you need a reliable Internet connection? Because the music isn't actually on your device. When you were listening to your own music, you literally transferred the music from your home computer, via a USB cable to your NOOK Color. With streaming music, your NOOK Color doesn't keep the songs on the device. Instead, like a radio, the device tunes into a station and just broadcasts whatever is playing. Unfortunately, that means that you can't listen to streaming music at, say, the beach unless you have a nearby Wi-Fi connection.

The second thing you'll need is a music-streaming app. NOOK Color doesn't have a built-in radio, so you can't just tune in to a particular signal like you would in your car. Luckily, NOOK Color has a wide array of apps that will find stations for you. Most of them are free, too. Here are two of the most popular ones:

▌ Napster

▌ Pandora

Napster is a revamp of the well-known music software from the late 1990s. Now a legal service, Napster has more than 12 million songs available for little to no money.

Pandora is a more recent app. The service asks you for information on your favorite music. It will then play a song for you, which you can say you like or dislike. The service keeps playing songs until it is able to detect a pattern and—voila!—it knows what music you like with shocking accuracy.

Barnes & Noble is a big fan of Pandora to the point where it is included on your NOOK Color. No need to visit the online App Shop to find it.

To get to Pandora, tap the NOOK button on the front of your NOOK, and tap the Apps icon in the Quick Nav Bar. You'll see the blue-and-silver Pandora icon in your app collection. Touch the icon to get started.

Pandora is initially a free service, but you do need to create an account. If you are already a Pandora user through the Web or another device, go ahead and type in your registered email address and password. If you aren't a user yet, tap the Create New Account option and it will ask you for the following info:

- Preferred email address
- New password
- Birth year
- Zip code
- Gender
- Join newsletter
- Agree to Terms and Conditions

The newsletter isn't necessary, but it can be helpful to get tips and tricks when you first get started. However, you do have to agree to Pandora's Terms and Conditions, which can be read by clicking on the live link under Terms and Conditions.

Once you put in the information, Pandora will ask you for a favorite song, artist, or composer. For instance, if you really like jazz trumpeter Miles Davis' classic album *Kind of Blue*, you can type in "Miles Davis" (artist) or "All Blues" (a song from *Kind of Blue*). Here's where it gets interesting: Pandora usually won't play anything from *Kind of Blue* or even necessarily anything from Miles Davis, but it will play music similar to or inspired by the song or artist. For instance, Pandora may play a song from Miles Davis' friend and fellow band member John Coltrane.

Now that a song is playing, you can help Pandora make your personal station better. The screen will show you the following info:

▶ Album cover
▶ Song length and information
▶ Name of song
▶ Name of artist
▶ Name of album
▶ Control icons
▶ Station icons

The song length may be self-explanatory, but try tapping the small "i" icon next to it. The album cover will flip and Pandora will tell you exactly why it chose to play the song you're listening to. Depending on your music knowledge, you may not know what "syncopated drums" or "challenging guitar riffs" mean, but Pandora recognizes such patterns in the songs you like.

The Control icons lie right beneath your album info:

▶ Thumbs Up
▶ Thumbs Down
▶ Bookmark

- ▶ Play
- ▶ Skip

Thumbs Up means you like this song, while Thumbs Down, of course, means you don't like it so much. These two simple icons are the key to Pandora. Thumbs Up a song and Pandora will

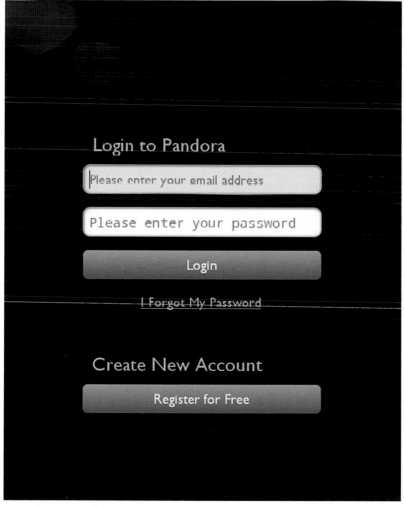

Pandora's sign-in screen

analyze the details about the song: the rhythm, the length, the vocals, and so on. It will then compare it to another song that you liked and see if there are any similarities. Now, the next song Pandora plays will, at minimum, have at least one quality shared between the previous songs that you liked.

Thumbs Down is essentially the reverse. If you don't like a song, Pandora will match it up to other tunes you gave a Thumbs Down to and see the similarities. It can be a lot trickier to avoid certain traits completely—like having a rock track with no guitar—but Pandora will try to minimize the incidence of traits you seem to dislike in its music selection.

Bookmark means you'd like to remember the song playing. Because of licensing restrictions, you can't rewind a song in Pandora or even play it again. In other words, Pandora isn't the same as your mobile music player or even your NOOK Color. What you can do is bookmark a song. Once it is bookmarked, Pandora will give you a link so you can buy it from a digital music distributor.

Play is just like your NOOK Color music player. Tap it to play the music or pause the music. You'll want to utilize this icon since you can't rewind or replay songs in Pandora.

Finally, the Skip icon will let you jump to the next song. Pandora will only let you skip a certain number of songs every 24 hours, so be judicious in choosing which song you absolutely want to skip. Worst-case scenario, you can turn the music down or remove your earphones until the song is done.

The station icons are just below your Control icons:

- My Stations
- Add Station
- Sign Out

Tap the My Stations icon and you'll get a list of your stations. Pandora allows you to create multiple stations, so you could have one dedicated to, say, sports anthems and another one focused on opera. Once you have a collection of stations, use your finger to scroll up and down the list. Tap a station to start it up. The Edit icon at the bottom allows you to delete a station: Touch Edit, then tap the "-" symbol next to any station to remove it.

Be careful with stations, as once you remove one, it's gone forever—and some stations have dozens, if not hundreds, of your ratings saved within them!

Add Station will allow you to create a new station based on a particular artist, song, or composer. It will be added to your My Stations list.

Lastly, Sign Out will log you out of Pandora. If multiple people use your NOOK Color, you can log out and allow others to use their own Pandora accounts on the device.

Watching Video

Once you master playing music, watching videos is even easier. Again, go back to the My Files folder. Tap on the Video folder and you'll see your videos listed.

Tap on your video and, if the format is compatible, it will start playing. You can use the built-in speaker or headphones, both of which can have their volume adjusted by using the two Up and Down buttons on the right side of your NOOK Color.

With your video on the screen, you'll immediately see four controls:

- Back (located in the upper left-hand corner)
- Rewind/Restart
- Play/Pause
- Fast Forward/Skip

Tap the Back button and NOOK Color will take you out of the video and back into the My Files section.

Rewind and Fast Forward will do just that when you hold them, but they will also restart or end the video if you tap the respective button. Finally, you can tap the Play button to pause, and tap it again to continue the video.

Video on NOOK Color

21. Apps

Apps are simply the software programs used by a device and downloaded directly onto your device from an online provider like the NOOK Store. The apps allow you to do tons of cool things on your NOOK Color, from finding recipes, getting news, or playing games.

NOOK Color has thousands of apps available right now. Most exceptionally, there are a large number of free apps, making it easier to take an app for a spin before you decide to keep it.

The categories are:
- Education & Reference
- Games
- Children
- Lifestyle & Interests
- Entertainment
- Themes
- Productivity
- Tools & Utilities
- Health and Fitness
- Social
- News & Weather

As you may imagine, Games and Productivity are two of the most popular app categories. We'll spend some significant time discussing those in chapters 22 and 23.

The Apps Screen

To get to the apps menu, tap the NOOK button on your device and open up the Quick Nav Bar. As you may remember, your menu will have the following options:

NOOK Color's main Apps screen

- ▶ Home
- ▶ Library
- ▶ Shop
- ▶ Search
- ▶ Apps
- ▶ Web
- ▶ Settings

You, of course, want to tap Apps. The Apps screen will now appear.

The Apps screen has a clean, but effective, design. We'll start from the middle. In the main section you'll see the apps you've downloaded. Already see a bunch of apps there? Barnes & Noble pre-loads your NOOK Color with some apps it thinks you'll love. In fact, while a couple are just fun and games, some of them are absolutely essential for you to get the very most out of your tablet. In this chapter we discuss the pre-installed apps in depth, but know that the center of this screen will change based on your tastes and needs.

Just above your collection of apps is a bar with a shopping bag icon and an arrow. The Discover Bar is the gateway to apps in the NOOK Store. We'll talk about all the cool shopping later in the chapter.

For now, look at the lower left-hand corner for the Sync icon (looks like two arrows making a circle). Apps are software, and the creators are always making updates to fix bugs, improve user experience, and even to add content. The frequency of these updates really depends on the app creator, but it's important to check regularly to make sure you have the latest version of an app. When you tap the Sync button, your NOOK Color will hop online and update the apps with their latest versions. You need to be online to sync your apps, so be sure to take advantage of it while you are in a Wi-Fi hotspot.

The Apps on Your NOOK Color

NOOK Color comes with a handful of cool apps pre-installed, so you can begin playing with apps immediately. As of Winter 2012, here's what you'll get for free on your app screen:

▶ Chess

▶ Contacts

▶ Crossword

▶ Email

▶ Music Player

▶ My Media

▶ NOOK Friends(TM)

▶ Pandora

▶ Sudoku

Chess and Crossword are digital versions of the classic board game and Sunday newspaper diversion, respectively. Sudoku is the popular number game. These are just a few of the games available for NOOK Color, which we'll get deeper into in Chapter 22.

My Media manages your pictures. Believe it or not, NOOK Color can also be a handy picture viewer, capable of full-screen displays, slideshows, and light photo editing. If you tap on the My Media icon, NOOK Color will show you all the pictures you have on file. NOOK Color comes with more than a dozen interesting photos pre-installed. Use your finger to scroll and you can coast up and down My Media. See a picture you like? The photos you're viewing now are called thumbnails, or miniature versions of the full-sized photos shrunk for the sake of navigating. Tap the photo, however, and you'll get the full picture in all its detail.

The Music Player app allows you to play songs transferred from your computer. We talked about it in Chapter 20, *Music and Video*, along with Pandora, the popular Internet radio service.

Contacts and email apps turn your NOOK Color into a mini-computer. In fact, you can take care of quick tasks just by using these two apps.

Finally, NOOK Friends enables you to connect with others who are also using NOOKs or NOOK eReader apps. You can look at it as a little social network for NOOK Color users. We'll go into detail about NOOK Friends in Chapter 24, *Getting Social*.

The Apps Shop

The handful of apps on your NOOK Color is a good start, but there are hundreds of other apps that are worth checking out.

Furthermore, there are dozens more coming out every week. The place to get them is the online store, which we touched on in Chapter 16, *NOOK Store*. Now we'll go further into the apps section.

Tap the "n" button, get the Quick Nav Bar, and tap Shop. Here you'll find the books, periodicals, and other items. Tap the Apps icon. Similar to the NOOK Store page, the Apps Shop shows the following parts:

- App Categories
- Top Pics in Apps
- Top Picks in Game Apps
- What's New in NOOK Apps
- Browse and Search Bar

Starting from the top, App Categories lists all the different types of software available. You can use your finger to scroll the list of catagories up and down. If you're looking for a specific type of app, you can tap one of these categories and NOOK Color will narrow down your search.

Below the app list are Top Picks in Apps, Tops Picks in Game Apps, and What's New in NOOK Apps. There are icons for the apps, but, unlike the vertical categories list, the three app lists run horizontally. Use your finger on any of the lists, and push it to the left or right to move it. You can also tap the See All option on any of the lists to see all of them listed vertically.

Go ahead and tap one of the app categories or the See All option on one of the lists. You'll see all the apps in that category. In the upper right-hand corner, you can choose your app layout from Picture, Picture and Text, and Text-Free. The Picture option makes the App icons big and limits the details to the title, company, rating, and price. The Picture and Text option has those same details, but also gives a brief synopsis of the app. Finally, the stripped-down Text-Free option removes the synopsis and shrinks down all the app icons, too. The Picture option emphasizes the visual, the Picture and Text option gives a good synopsis, and the Text-Free option puts as many app choices as possible on one screen. Play around with them and decide which one works best for you.

Once you choose the type of layout you want, take a good look at the apps selection. Depending on the layout you choose, the apps will have the following details:

- Picture
- Title
- Company

▶ Rating

▶ Price and Purchase button

The picture is the logo or image for the app. You'll see this picture on your apps screen after you download the software, and it will be what you click on to start the app. The title and company are the name and the creator of the app.

The rating is how users have ranked the app. NOOK Color uses a five-star system. You'll notice a number in parentheses right after the star rating. It represents the number of people who have rated the app. Why is it important? The more people who have ranked the app, the more likely that the ranking will be representative of the app's quality. For example, if only one person ranked an app, that person may have a certain extreme bias. By telling you how many people have rated it, NOOK Color makes it easy to determine how much weight you should put on the current ranking.

The Price and Purchase button gives you the current cost and the option to download the app. If you decide to buy the app, the money is automatically deducted from the credit card you submitted when you first registered your NOOK Color. We'll have some fun downloading apps in the next section.

Finally, check out the Search Bar at the bottom of your touchscreen. Tap the icon and, near the bottom, tap Apps.

Let's take a look at the Search Bar. Tap the Search section and it will open up the Search page.

Go ahead and type in the name of an app with the keyboard. Let's try Aquarium Live Wallpaper, a Bestselling app that turns your NOOK

Color screensaver into a lifelike fish tank. You'll notice that, as you type, the Suggestion Box will fill up with several different apps that fit the name. You can now tap one of the suggestions and open up that particular app in the Apps Shop.

Trying to remember something you searched for a while ago? Once you do a search, NOOK Color will remember the search terms you used. These are shown before the search suggestions so you can pick up where you left off during a previous search.

Downloading Apps

Now that you know how to browse, why don't you find an app you like? When you find an interesting app, tap on the icon. You'll get a detailed overview of the app with this info:

- Name
- Company
- Version
- Rating
- Price and Purchase icon
- Add to Wishlist
- Share
- Overview
- Customer Reviews
- Screen Shots
- More Like This

The Name, Company, Rating, and Price and Purchase icons give the details we talked about earlier in the chapter, while Version tells you which version number is available for download. As app creators update the software, they will increase the version

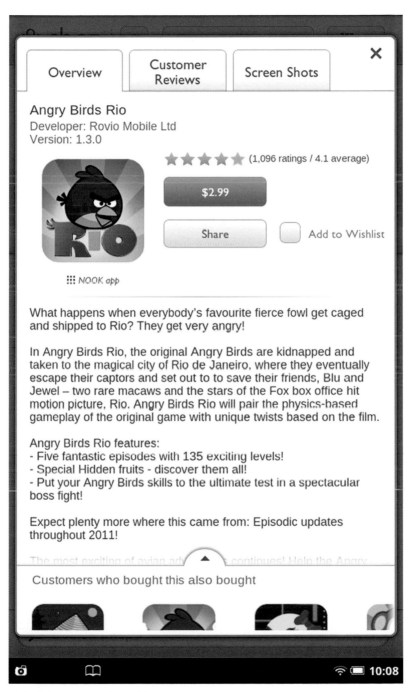

Overview | Customer Reviews | Screen Shots

Angry Birds Rio
Developer: Rovio Mobile Ltd
Version: 1.3.0

★★★★★ (1,096 ratings / 4.1 average)

$2.99

Share   Add to Wishlist

⫶⫶ *NOOK app*

What happens when everybody's favourite fierce fowl get caged and shipped to Rio? They get very angry!

In Angry Birds Rio, the original Angry Birds are kidnapped and taken to the magical city of Rio de Janeiro, where they eventually escape their captors and set out to to save their friends, Blu and Jewel – two rare macaws and the stars of the Fox box office hit motion picture, Rio. Angry Birds Rio will pair the physics-based gameplay of the original game with unique twists based on the film.

Angry Birds Rio features:
- Five fantastic episodes with 135 exciting levels!
- Special Hidden fruits - discover them all!
- Put your Angry Birds skills to the ultimate test in a spectacular boss fight!

Expect plenty more where this came from: Episodic updates throughout 2011!

The most exciting of avian adv͟͟͟ continues! Help the Angry

Customers who bought this also bought

10:08

Detailed overview of an app

number. Version 1.0 means it is the very first version released to the NOOK Store, and incremental numbers mean revisions. If you sync regularly, you'll always have the latest version of an app. Some apps have two additional options: Add to Wishlist and Share. Add to Wishlist will put the app on the record as something you want. Your Wishlist can be kept for your own memory or shared with friends, as we'll discuss in Chapter 24, *Getting Social*. Share allows you to tell your friends about the app via your NOOK contacts, through Facebook, or on Twitter.

The remaining details of Overview, Customer Reviews, and Screen Shots, are at the top of the screen, and More Like This is available at the bottom. Tap the tab to see the info you'd like.

Overview gives you a longer description of the app. It usually is a long paragraph written by the app creator. It can also include more info on the creator, including its Web address and contact info.

Customer Reviews lists all the reviews of the app with the review title, the user name, the rating, the posting date, and, of course, the review itself. Use your finger to scroll up and down the list of reviews. NOOK Color lists the ten most recent reviews, but additional reviews can be read by touching the More icon at the bottom of the listed reviews.

Also, if you want to share your opinion on an app you've downloaded, you can tap the Write a Review icon found in the upper left-hand corner of the review list. The Rate and Review page asks you to give an overall rating, a headline, and a review. Don't worry about space; there's room for up to a 3,500-character review. We're assuming you haven't used the app yet, so you can touch the Cancel button. When you're ready to write a real review, touch Post to upload it to the NOOK Store.

The Screen Shots area gives you an idea of what the app will look like on your NOOK Color before you download it. There are usually several shots, so you can use your finger to scroll down and view them all.

Lastly, the More Like This section recommends apps that are similar to the featured app. For instance, if you're looking at a calculator app, the Apps Shop might suggest apps related to finance or geometry.

It's fine if you decide against buying an app, too. Look in the upper right-hand corner and you'll notice an "X." Tap it to close the detailed app view and get back to your Apps Shop list.

However, if you do want to buy the app, tap the Price and Purchase icon. Whether it costs money or not, NOOK Color will ask you to confirm the purchase. You'll move through the following steps:

- Confirm
- Purchasing
- Downloading
- Installing
- Open

If your Wi-Fi is solid, the Purchase and Confirm steps will take just a few seconds. Downloading time depends on your Wi-Fi speed and the size of the app. For instance, a simple calculator app could take a few seconds, while a robust visual app might take several minutes. As it downloads, a green bar will appear on top of the app's icon. Your app is downloaded when the bar is full.

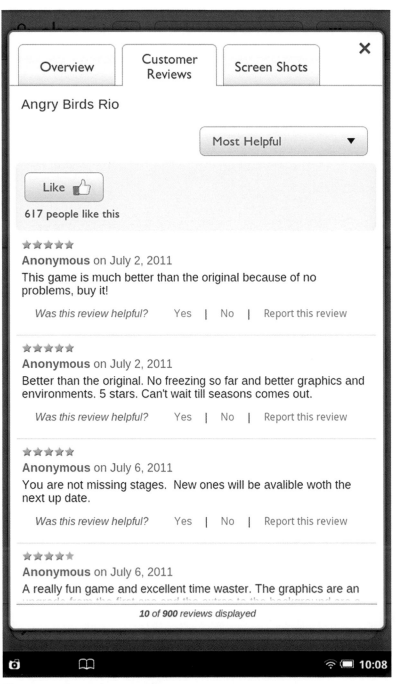

| Overview | Customer Reviews | Screen Shots | ✕ |

Angry Birds Rio

Most Helpful ▾

Like 👍

617 people like this

★★★★★
Anonymous on July 2, 2011
This game is much better than the original because of no problems, buy it!

Was this review helpful?   Yes | No | Report this review

★★★★★
Anonymous on July 2, 2011
Better than the original. No freezing so far and better graphics and environments. 5 stars. Can't wait till seasons comes out.

Was this review helpful?   Yes | No | Report this review

★★★★★
Anonymous on July 6, 2011
You are not missing stages. New ones will be avalible woth the next up date.

Was this review helpful?   Yes | No | Report this review

★★★★★
Anonymous on July 6, 2011
A really fun game and excellent time waster. The graphics are an

10 of 900 reviews displayed

📷   📖   🛜 🔋 10:08

The rate and review app options

Next, NOOK Color will install your app. Again, the installation time will depend on the size of the app you downloaded. It should take no more than a couple of minutes.

Once the app has finished downloading, the button will read Open. Tap the button and your app will start up. When you're on the app's main page, you can just tap the icon to start up the app.

Managing Apps

As you explore more apps in the Apps Shop, you'll find several that will intrigue you and others that just won't be as interesting. Some, as we mentioned earlier, will be useful, but will take up more memory than you'd like or are not needed frequently. Regardless of the reasons, you'll want to manage your apps.

First, go to your main Apps screen. Here you can tap an app to start, but try holding your finger down on an icon of an app you've downloaded. You'll get the following menu:

- Open
- View Details
- Recommend
- Add to Home
- Add to Shelf
- Archive
- Delete

Open will start the app for you. View Details and Recommend will give you the info found on the app's Purchasing page and allow you to tell friends about the app.

The three most important options here are Add to Home, Add to Shelf, and Archive. Add to Home lets you move a link to the app on to your NOOK Home Screen. Say you use a calculator app a lot, so you'd prefer not to have to start up your NOOK Color, pull up the menu, tap the Apps option, and finally tap the calculator icon every time to use it. If you choose to add the icon to the your Home Screen, NOOK Color will create a quick link to your app. Turn on NOOK Color and you'll see the app sitting at the bottom of the screen. Remember that it is just another link to your app, so your app can still be found on your actual Apps screen, too.

Add to Shelf will move the app to a specific virtual shelf on your NOOK Color. It will ask you which shelf you'd like to move it to or the name of the new shelf you'd like to create. Now you can organize your apps using your own system.

However, if you do want to remove the app from your NOOK Color, then the Archive option is perfect. As we touched on earlier, archiving takes an app off your NOOK Color and puts a link to it on the Archived page. Archiving an app can save you memory, because the item is removed from NOOK Color, and it can keep your app page from getting too cluttered. If you'd like to get an app back onto your NOOK Color, you need to select Archived from the drop-down menu under the My Stuff icon in the Library. Choose the app you'd like to retrieve and tap the button labeled Unarchive. You can now get the app and go through the download process again. Keep in mind that NOOK Color will need Internet access to re-download your apps from the archive. The final option, Delete, will permanently remove apps that you have no interest in downloading again.

Don't worry about downloading games just yet, as NOOK Color comes with some games pre-installed. Your NOOK Color doesn't differentiate between apps and games, so your game collection will always be on your App screen.

> Apps that are found to be appropriate for kids will be on both the Apps screen and the Kids screen.

To get to the App menu, tap the NOOK button on your device and open up the Quick Nav Bar. Touch the Apps icon. The Apps screen will now appear.

Top Apps

Not sure where to start? It can be pretty overwhelming to decide which apps you should download first. NOOK Color has a ton of choices, but certain apps are perennial favorites on the device. The apps are found in the Apps area of the NOOK Store. Here are some high-rated best sellers:

Bling My Screen • *Murtha Design*

NOOK Color has plenty of modification options, but extra choices are always welcome. Bling My Screen helps you personalize your Home Screen with additional wallpapers, virtual bookshelves, and other organizational tools.

Fandango Movies • *Fandango*

Like movies? The Fandango app makes it easy to find out what movies are playing, learn about the films, and even purchase tickets for tonight's show. It also gives lots of theater-buff detail, from critics' reviews to box-office sales.

22. Games

As we discussed, NOOK Color comes with a handful of pre-loaded games:

- Chess
- Sudoku
- Crossword

Chess and Crossword are digital updates of the classic board game and Sunday newspaper diversion, respectively. Sudoku is a popular number game.

Understanding Controls

NOOK Color doesn't have any buttons aside from the NOOK button. Its power lies in the flexibility of the touchscreen itself.

When it comes to games, your NOOK Color uses a few different control techniques:

- Tap
- Double tap
- Slide
- Hold

Tap is the method you've been using to do different things with your NOOK Color so far. It simply means touching the screen. You tap icons, books, and even menu choices to open them.

Double Tap means tapping the touchscreen twice in rapid succession. It isn't necessary for most actions, but you'll find that some games will require it.

Slide is touching a part of the touchscreen and dragging your finger to another part of the touchscreen. A slide can move a game character from one place to another, map the direction of a toss in a game, or handle several other actions.

Finally, Hold is doing a tap, and then keeping your finger down. It requires having a steady hand, as NOOK Color can sometimes think that you are sliding instead of holding.

It might help to see the touchscreen moves in action, so why don't we take a game for a test drive. Tap on the Chess icon.

If you know chess, everything will look familiar to you: 16 pieces on each side of an 8-by-8-square board and a timer to see how long each player is taking with his or her turn. Pretty basic, but how do you actually play chess on NOOK Color?

As you are learning, the touchscreen can create virtual, touchable buttons anywhere on it. Tap the New Game icon and a new game starts. Touch the Settings button and you can modify your color, difficulty level, and time limits. Once you get a new game started, try touching one of your pawns—one of the eight short pieces located in the second-to-bottom row on the board. Hold your finger on it, then drag your finger to the square above it. Look carefully and you'll see your pawn following your finger. Let go of the screen and, assuming it is a legal move, the pawn will move to the new square. You just slid a piece to where you wanted it to be and completed a move.

One more example might help here. Tap the NOOK button to get the Quick Nav Bar, touch the Apps icon, and tap the Crossword icon.

Again, the game looks like your traditional Sunday newspaper crossword: a big, blank graph with clues for the answers going across and down. Tap on a square, and the clues for the corresponding across and down answers will appear at the top.

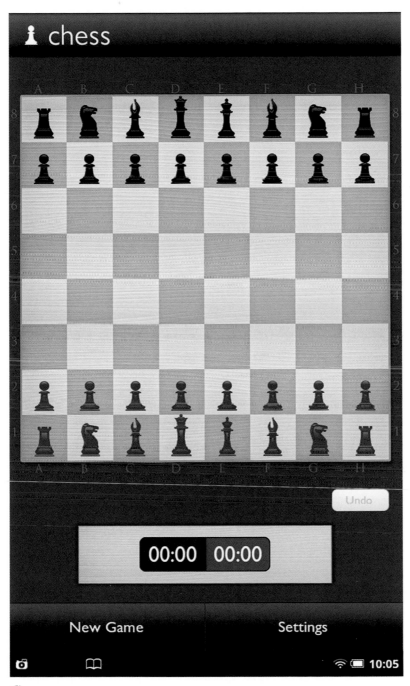

Chess

Notice how one line, either across or down, is highlighted? If you start typing in an answer, the game will automatically begin filling in that line. However, what if the line is going across and you want to go down, or vice versa? Double tap the square and the game will flip the highlight vertically or horizontally. In other words, tap once to put your focus on a particular square, but double tap to highlight surrounding line either across or down.

Like apps, each game has its own particular set of controls. You can check the Help or Settings page of each game for details. Or, if you are feeling adventurous, play around with different touches and find out what they do. You can always reset the game!

The Games Screen

The last chapter described the Apps screen, which is where your games will be stored. To get to the Apps menu, just tap Apps in the Quick Nav Bar, and you'll be taken to the Apps screen in the Library.

Shop Now

Before you go on a shopping spree, check out the lower left-hand corner for the Sync icon (looks like two arrows making a circle). Sync lets NOOK Color check to see if there are any updates to your apps, including your games. Software can have bugs, like little glitches that make the game less enjoyable than it should be. And with games, sometimes companies will add free content just to keep you playing! Either way, game companies will give you a heads-up when a new, improved version of your game is out. Tap the Sync icon and your NOOK Color will download whatever updates are available for the games you own. Like archiving, syncing requires a Wi-Fi connection. Download times can vary, but, to be on the safe side, use the Sync option when you know you'll have more than a few minutes. In fact, your local Barnes & Noble store has free Wi-Fi, so you can always

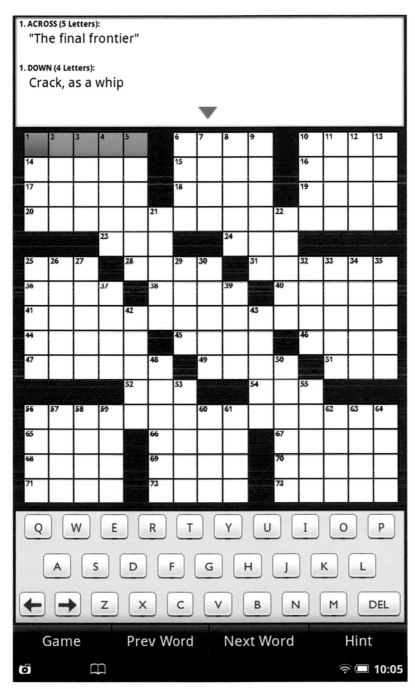

Crossword

Categories of apps for sale

visit, sync up your NOOK Color and browse the NOOK Store while the device downloads the latest versions of your favorite games.

The Apps Shop

Ready to get your hands on some new games? Like apps, your entry to games is through the Apps area of the NOOK Store.

Get back to the Apps Shops by touching the "n" button, selecting Shop, and then Apps. Similar to the books area in the NOOK Store, the Apps Shop shows the following parts:

- Order and layout
- App Details
- Browse and Search Bar

As discussed in Chapter 21, *Apps,* the Order and Layout icons will change the format of the list of apps, Apps Details gives info on the highlighted app, and the Browse and Search Bar lets you type in a particular term to find the game you want.

Each game will have the Picture, Title, Company, Current Rating, and the Price and Buy Now button.

Now tap the Apps icon under the NOOK Store. It will give you a list of categories, from Children to Tools & Utilities. We want Games, so tap on that category. There are thousands of apps, but now NOOK Color will only show you the games.

There are now three areas:

- Genre
- Bestselling
- Search

Shopping for games

Genre will only show you games of certain categories, like Kids or Arcade. Bestselling lists the hottest games of the day.

Finally, searching is the most powerful tool you have in the Apps Shop for finding games. Tap the Search Bar and you'll get the following options:

- Search Suggestions/Recent Searches
- Search Bar
- Keyboard

Go ahead and type in the name of a game with the keyboard. Let's try the name of the super-popular game Angry Birds. You'll notice that, as you type, the suggestion box will fill up with several different apps that fit the name. You can now tap one of the suggestions and open up that particular app in the Apps Shop.

Trying to remember something you searched for a while ago? Once you do a search, NOOK Color will remember the search terms you used. These are shown before the search suggestions pop up when you begin typing in a term. It is a quick, simple way to pick up where you left off during a previous search.

Downloading Games

Let's start downloading games. Find a game you like and, as with other apps, you'll get details, including:

- Name
- Company
- Version
- Overview
- Customer Reviews

- Rating
- Screen Shots
- Price & Purchase
- Add to Wishlist
- Share
- More Like this

Many of these details were discussed in the last chapter, but the most important ones are Price, Customer Reviews, and Screen Shots, so you can see how the game looks. The Recommendations section, which tells you "Customers who bought this also bought..." can be helpful after you buy the game because, if you like it, you can find out about other similar apps.

If you decide not to buy a game, you can touch the "X" in the upper right-hand corner. This will bring you back to the general games list.

When you find a game you would like to buy, touch the Purchase icon. Your NOOK Color will ask you to confirm the purchase. To approve it, you'll use the following process:

- Confirm
- Purchase
- Download
- Install
- Open

Purchase and Confirm should just take a few moments. Downloading the game takes a little while longer, depending on your Internet connection and the memory size of the game. Big, complicated games take longer to download. You'll see a green bar on top of the Game icon that tracks how much has downloaded so

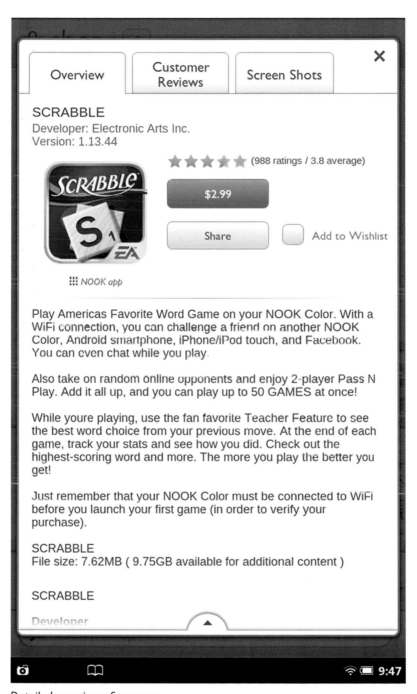

SCRABBLE
Developer: Electronic Arts Inc.
Version: 1.13.44

★ ★ ★ ★ ★ (988 ratings / 3.8 average)

$2.99

Share   Add to Wishlist

⋮⋮⋮ NOOK app

Play Americas Favorite Word Game on your NOOK Color. With a WiFi connection, you can challenge a friend on another NOOK Color, Android smartphone, iPhone/iPod touch, and Facebook. You can even chat while you play.

Also take on random online opponents and enjoy 2-player Pass N Play. Add it all up, and you can play up to 50 GAMES at once!

While youre playing, use the fan favorite Teacher Feature to see the best word choice from your previous move. At the end of each game, track your stats and see how you did. Check out the highest-scoring word and more. The more you play the better you get!

Just remember that your NOOK Color must be connected to WiFi before you launch your first game (in order to verify your purchase).

SCRABBLE
File size: 7.62MB (9.75GB available for additional content)

SCRABBLE

Developer

9:47

Detailed overview of a game

far. Stay in your Wi-Fi area while the game is downloading, or the process may be interrupted.

Once it finishes downloading, the game will automatically install on your NOOK Color. Finally, when the Open button appears, you're ready to play! Tap it and the game will start. The game is also installed on your main Apps screen. You can always tap the icon from there, too.

Managing Games

As was mentioned earlier, some games will take up a lot of memory in your NOOK Color. In fact, games in general tend to be bigger in size, so the chances of you wanting to archive the latest NOOK Color game to save on memory are a lot higher than, say, needing to archive a simple calculator app. As a reminder, archiving means taking an app off your NOOK Color but keeping it available, via Wi-Fi, for re-download whenever you'd like to play it again.

To organize your games, go to the main Apps screen. Hold your finger on any game icon. You'll see a menu pop up:

- Open
- View Details
- Recommend
- Add to Home
- Add to Shelf
- Archive
- Delete

Top Games

NOOK Color has a serious collection of games for both joystick jockeys and casual gamers. Here are some favorites:

Angry Birds • *Rovio*

One of the most popular mobile games ever made, Angry Birds has you tossing birds to take down the structures of the greedy pigs. Underneath the cartoon atmosphere is a super challenging (and addictive) puzzle game that teaches physics.

Solitaire • *Agile Fusion*

A solid version of the classic card game, Solitaire pushes you to put random cards in order as quickly and as smartly as possible. Peppered with relaxing music and nice graphics, Solitaire has different game modes to keep things challenging.

Doodle Jump Deluxe • *GameHouse*

Doodle Jump was one of the first major hit tablet games, and NOOK Color's update keeps the spirit of the original. Simply slide your finger or tilt the device to bounce from platform to platform. The higher you go, the bigger the points and the more worlds you can explore.

Scrabble • *Electronic Arts*

The word battle Scrabble has already been a classic in board game form, but NOOK Color's version adds plenty of new features. The hint system and tough computer opponent are great. The biggest benefit, though, is being able to play against your friends online competing for the best word score.

Open, View Details, and Recommend are fairly straightforward. Add to Home lets you put the icon on your NOOK Color Home Screen. Play one game a lot? You may want to put it on your Home Screen so you have the quickest access to it, whenever you want to play your game.

Archiving is the key to keeping your game (and app) collection tidy and orderly. While you're on this menu screen, tap Archive, and NOOK Color will take the game off your device and move it into online storage. Whenever you want the game back on your NOOK Color, tap the Archived icon, which appears in the drop-down menu that opens after you tap My Stuff in the Library, and you'll get a list of all the apps you've archived. Remember that you need a solid Internet connection to unarchive any games. You can also permanently remove a game by selecting Delete.

23.
Productivity

We've discussed fun apps and gaming in the previous chapters, but what if you actually want to get work done on your NOOK Color? The virtual keyboard, copious memory, and easy portability make that easy to do, using the dozens of productivity apps in the NOOK Store.

The Productivity Screen

Like other apps, all your productivity software is on the Apps screen in the Library. To get to the Apps Shop, touch the "n" button to pull up the Quick Nav Bar, tap Shop, and then the Apps icon. In the genre box at the top, scroll the catagories until you see Productivity.

To find a certain Productivity app, you'll want to use the Search feature. Tap the Search Bar and you'll find the following options:

▶ Recent Searches
▶ Search Bar
▶ Keyboard

Using the keyboard, type in "Quickoffice," the name of the popular word-processing suite. You'll notice that, as you type, the suggestion box will fill up with several different apps that fit the name. You can now tap one of the suggestions and open up that particular app from the NOOK Store.

Trying to remember something you searched for a while ago? NOOK Color will remember the search terms you used. These are shown before the search suggestions pop up when you begin typing in a term. It is a quick, simple way to pick up where you left off during a previous search.

Productivity apps

Downloading Productivity Apps

Like all apps, your office apps will give you the following:

- Name
- Company
- Version
- Rating
- Price and Purchase icon
- Add to Wishlist
- Customer Reviews
- Share
- Screen Shots
- Overview
- More Like This

Pay attention to Price, Reviews, and Screen Shots before you purchase. If you don't want to buy an app, tap the "X" in the upper right-hand corner to get back to the main apps list.

When you find a productivity app you want to buy, tap the Buy Now button to begin the process:

- Confirm
- Purchase
- Download
- Install
- Open

Purchase and Confirm are quick, but downloading the actual program can take a few minutes. Watch the green bar that appears on the app icon to see how long you have to wait. Remain in a Wi-Fi hotspot to make sure the download isn't interrupted.

After it downloads, your NOOK Color will install the app. The Open button appears once the app is ready. Tap it to start your app. Remember, you can always run your app from the Apps screen in the Library too, as your icon will appear there.

Managing Apps

Check out Chapter 21, *Apps* for more details on organizing apps, but know that holding your finger on the Apps icon will give you a menu with the following options:

▶ Open

▶ View Details

▶ Recommend

▶ Add to Home

▶ Add to Shelf

▶ Archive

▶ Delete

Top Productivity Apps

There are many apps that will help transform your NOOK Color into a full-blown tablet. Check out this software to get work done on the go:

Quickoffice Pro • Quickoffice

Quickoffice Pro lets you stay on top of work commitments while you're on the go. The office suite is compatible with Microsoft Office Word, Excel, and PowerPoint, and it syncs up to your documents through Google Docs, Dropbox, and other software.

Calculator & TipCalc • 5ivedom

This low-priced app features a built-in tip calculator. The main calculator, capable of sine, cos, and tan measurements, isn't too bad, either.

My-Cast Weather Radar • Garmin Digital Cyclone

NOOK Color doesn't come with a weather app, and this handy one from GPS leader Garmin fits the bill. Download My-Cast® for hourly or extended forecasts, detailed maps and graphs, or international weather information for your travels.

24. Getting Social

Not only is NOOK Color light and portable, but the Internet connection makes it easier to make new friends online, reconnect with old friends, or even share your books with other people. All this can be found in NOOK Friends.

To start up NOOK Friends, press the NOOK button to open the Quick Nav Bar. Tap Apps and then NOOK Friends on the Apps screen.

There are three icons under the NOOK Friends app:

- Friends' Activities
- About Me
- LendMe

The default page, Friends' Activities, gives you a news feed telling you what your NOOK Friends are doing. Your friend Sarah might have just recommended the latest Mary Roach book, while Paul just reviewed his latest sci-fi read. It's a good, simple way to stay connected to your fellow readers. You can look at it as a virtual book club.

NOOK Friends lists all your current reading-circle friends. You can also request friendships and send messages to current friends. We'll get into that in a moment.

The About Me page is the personal profile people see when they are your friends. It has a few facts about you, including the number of friends you have and the books you currently own.

Finally, LendMe tracks all the books that you've lent and borrowed, as well as titles you've asked to borrow and books people are willing to lend.

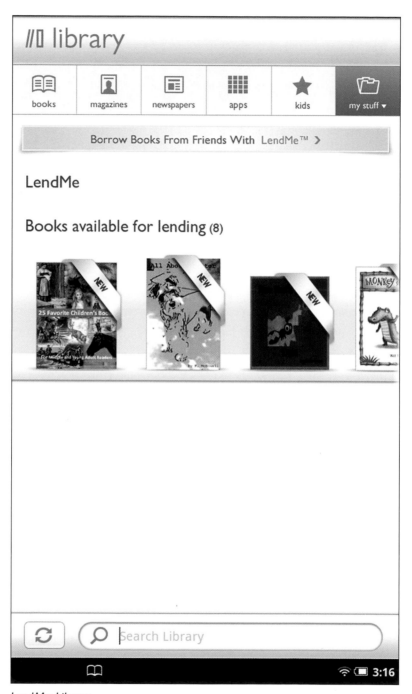

LendMe Library

If you own any books that you don't want others to see, there is a way to hide these titles from view. Learn more about this later in this chapter.

Finding and Making **NOOK** Friends

Tap the All Friends icon to see all your NOOK Friends. You'll see three buttons:

- Add Friend
- All Friends
- Pending

Let's actually start with the second icon, All Friends. Tap the All Friends button and you'll see a list of your current friends with their name, picture, and other stats. You haven't connected with anyone yet, so this section will be blank. The next section, Pending, shows all the people interested in being friends with you and, conversely, all the folks you've asked to be your friend but haven't responded yet. If someone has asked to be your friend, you can tap Accept (to become his or her friend) or Ignore (to make the request disappear). Don't worry: Like other social networks, NOOK Friends won't send a notice to people you choose to ignore. No need to feel guilty!

You can also drop NOOK Friends, too, by tapping the "X" icon just to the right of their name. Finally, you can borrow a book from them by tapping the LendMe icon and finding out what books they have available to lend to you.

Now, let's start getting some friends. Tap the Add Friend icon and NOOK Color will give you four ways to find friends:

NOOK Friends

▶ Find Friend from My Contacts

▶ Find Friends from Facebook

▶ Find Friends from Google

▶ Invite a Friend via Email

If you already know one of your friends is a NOOK user registered on BN.com (Barnes & Noble's Web site), you can tap Add New to contact him or her directly. NOOK Color will ask for the first name, last name, and email. Barnes & Noble will send your friend an email.

Suggested Contact is cool because you can also contact friends who don't have a NOOK yet, but may get one in the future. You will still go through the Add New process, but your NOOK Color will store the person's email address and give you a heads-up when he or she registers on the BN.com Web site. Your contacts who are newly registered BN.com members will be found under the Suggested icon.

Finally, you can have NOOK Color go through your social media contacts and see if any of your friends are NOOK users registered on BN.com. Tap the All Contacts icon, then the Set Up Account icon. You'll find three different ways to find new NOOK Friends:

▶ Facebook

▶ Twitter

▶ Gmail

You can choose the social media where most of your friends are listed. For instance, if you're a hard-core Facebook user but never touch Twitter, you can just link your Facebook account. There may be more social media options in the future, but Facebook, Twitter, and Gmail cover literally millions upon millions of users, all of whom are potential NOOK Friends.

Facebook is the easiest way to make connections because, once you sign in, it will automatically make you NOOK Friends with your Facebook friends. On the other hand, Twitter won't allow NOOK Color to search your friends lists for NOOK users, but it does make it easy for you to post any new book, app, or review onto that social network.

Connecting your NOOK Color to Facebook or Twitter is the same process: Tap Link Your Account, then type in your email and password. Your NOOK Color will confirm that you're allowing it to post onto your Facebook wall or Twitter feed. Now you can update friends about your NOOK Color activity and let them know that they can "friend" you through NOOK Friends.

If you have a Google account, and particularly a Gmail address, you can have your NOOK Color search your contacts and find any current friends on NOOK. (Luckily, Gmail is free. You can get your own account at www.gmail.com.) Like Twitter and Facebook, Gmail will ask for your email, password, and permission to connect to NOOK. Now any current registered NOOK users on your Gmail contact list will appear in the Suggested area. Because NOOK Color has to sort through the contacts, they usually take a few minutes to appear on the list.

Lending and Borrowing Through LendMe

Sharing good books with friends is one of the pleasures of physical books; you can lend a worn paperback to a friend and she can return it to you when she's done. If your friend happens to lose the book, it can be easily replaced.

You can't just hand off your NOOK Color to a friend so he can read one book; you'd be handing off your whole Library! Also, replacing a lost or damaged NOOK Color costs a bit more than the price of a paperback.

Luckily, your NOOK Color makes it easy to share your books with other NOOK users. They can borrow as many books as they like, and you, in turn, can borrow books from them, too.

The NOOK lending and borrowing program is called LendMe. To see your LendMe section, tap the LendMe icon on your NOOK Friends page. It is the third button.

LendMe has four sections:
- **Lend:** my lendable books
- **Borrow:** friends' books to borrow
- **Offers:** books friends will lend
- **Requests:** books others want to borrow

Lend shows the books in your collection that you can let others borrow. Like a physical book, your NOOK books can only be lent out to one person at a time and, on the NOOK Color, you can only lend a book once. Your friend has a week to take you up on your lending offer—otherwise the book comes back to you. If they do decide to borrow your book, they keep it for 14 days, a period during which you won't be able to read it. Keep in mind that some publishers and authors don't make their books available for lending, so NOOK Color will only show those books that can be shared.

Borrow shows all the books your friends have that you can borrow. You can request a book to borrow here.

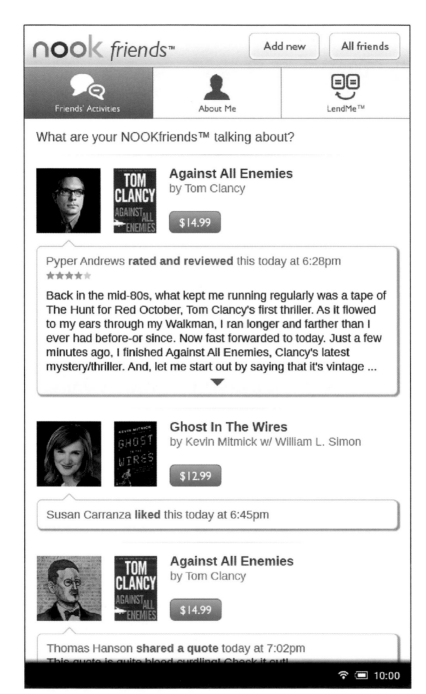

Comments from your NOOK Friends

Offers are books that your friends say they would be happy to let
you borrow. This can come in handy. Let's say you are having coffee
with a friend and she mentions a book she recently read that you'd
absolutely love, but she can't remember the title. Once she gets to her
NOOK, she can look up the book and offer it to you to borrow. Finally,
Requests are books your friends would like to borrow from your
current Library. You can let them borrow your book from here.

> Want to keep some of your reading material private? In the
> NOOK Friends section, tap the Privacy button in the upper right-
> hand corner. You'll now see a list of all your books. Toggle the
> Show icon to make a book visible or invisible to your friends.

Using Email

Another useful feature of your NOOK Color is email. It is located
on your Apps screen, which you can reach by pressing the NOOK
button.

NOOK Color will ask you for your email address and password,
the type of email (POP or IMAP), and other details specific to your
email. If you're not sure how to fill out the details, go with the
default information. If it doesn't work, visit the Help page of your
email Web site, and it should tell you how to allow devices to have
access to your email.

Once your email is set, you're ready to access your messages. Keep
in mind that NOOK Color doesn't create a new email account for
you, but just allows you to read and send email through your current
accounts, like Gmail or Yahoo.

25. Memory and Storage

All digital gadgets have a limited amount of memory, and over time the memory gets used up. NOOK Color has a lot of space for books—typically up to about 5,000—but the Library can get cluttered, especially when you start downloading apps, music, magazines, and newspapers. It's worth knowing how much memory you have, and how you can help mitigate digital overflow.

In order to see how much space is available for more books, periodicals, and apps, access the Settings menu by pressing the NOOK button to open the Quick Nav Bar. Then tap Settings. Once there, tap the tab that says Device Info.

The Device Info pane shows how much battery power is left, and how much memory is left. There are different configurations of NOOK Color and, based on your model, you will have memory allocated for system software and an even larger amount of memory available for content from the NOOK Store. You can go to your Advanced Settings and look under Device to see how much memory you have available. That space can hold countless books, apps, or anything else, but it's important to know how to manage and delete items just in case NOOK Color begins to run out of storage space.

In order to see how large a specific book is, simply double tap that book in the Library. The screen will tell you where it is stored and how large it is. The page will also give you information such as the title, author, and when the file was last modified.

Managing Space

Now that you know how much storage space is left, it's time to learn how to manage it. There are three different places to store books:

NOOK Color's Device Info screen

- In NOOK Color's memory
- In the Barnes & Noble Archive online
- On a microSD card

Archiving Items

To archive an item, go to the book's detail page by double tapping the book's cover in the Library, and tapping Archive. You should receive a confirmation message asking if you really want to do this. Tap Yes, and the book will be removed from your NOOK Color and stored in your account on Barnes & Noble's Web site. You can find it under B&N Lifetime Library.

You can only archive books you've purchased from the Barnes & Noble store. Anything that you've put on your NOOK Color from your computer can be added and removed at your own discretion. We'll be discussing how to connect your NOOK Color to your computer later in this chapter.

The Barnes & Noble Archive

The Barnes & Noble Archive is a list of books that aren't currently on your NOOK Color but are available for you to re-download. To access the Archive, go to BN.com and log into your account. Once there, move your mouse over the My Account tab and click My Library. On the left pane, there will be a button to view your Archive. In the Archive, you can delete anything you no longer want. The Barnes & Noble Library also lists what samples you have downloaded, and what books have been purchased in full.

The Barnes & Noble Archive

Re-downloading Items

Once your item is archived, it's extraordinarily easy to get a book back onto your NOOK Color. To do so, make sure you have a Wi-Fi connection. Then, go to the My Stuff section and tap the Archived option. Tap the book you want to unarchive, and the book starts to download back onto your NOOK Color. It's a one-tap process, and the book should be in your Library in a matter of seconds.

Your NOOK Color contains both a Quick Start Guide and a User's Guide. These cannot be archived because they are there to answer any questions you might have, even as you are using your device.

MicroSD Cards

The final way to manage space on your NOOK Color is with microSD cards, which are sold at most electronics stores and can be used for storing an additional 32 GB of books and personal files on NOOK Color. In order to use one, you first need to install the microSD card. To do so, turn off the device and insert the card in the slot by your NOOK Color's curved metal bar.

Now that the card is installed, you can load files onto it from your Library. In Library, go to the My Files tab in the upper left corner, tap it, and look for the button labeled Memory Card. Here, you'll be able to find all the PDF and eReader files on the card that NOOK Color can read.

If you want to transfer files to NOOK Color you'll have to connect NOOK Color to your computer.

Connecting to a Computer

Connecting NOOK Color to a computer is easy. Just take the USB-to-micro-USB cable that came with the device, plug the micro-USB end into NOOK Color and the USB end into the computer. Now the two are connected.

From here, you can manually add files to NOOK Color. To do so, simply right-click and select Copy for the files you want to put on the NOOK Color, navigate through the file named MyNOOK and paste them in. It doesn't matter what subfiles they're in, as long as they're in a supported format and are stored on NOOK Color. From there, the files should be accessible from the library.

> For this section, we're talking exclusively about how to connect NOOK Color to a PC or Mac. The process could be different for a Linux system.

By connecting your NOOK Color to a computer, you'll be able to add free eBooks from Web sites, such as the Gutenberg Project; which we'll be discussing later in the appendix.

eReader files that NOOK Color can read:

- ▶ ePub
- ▶ PDF

Wishlists

If you want to make a note to purchase something later, then Wishlists are the way to go. On the page for every book, right next to the book's star rating, there's a little box labeled My Wishlist. Tap the box to put a check mark in it. Every book you check mark will be on your Wishlist. The Wishlist is accessible by tapping the My Account button in the upper right corner of the NOOK Store. This opens a pull-down menu. Tap My Wishlist. With the Wishlist, you shouldn't have any trouble remembering which books to check out.

Managing space is paramount once your NOOK Color starts overflowing. You'll be able to fit hundreds of books in the given space of 250 MB, alongside the 750 MB for the B&N Store. Still, knowing how to sync your NOOK Color and archive books is key to keeping an organized Library.

26. Advanced Techniques

NOOK Color is filled with outstanding features and capabilities, many of which take you well beyond the simple enjoyment of reading. There are ways to create a personalized wallpaper for your NOOK Color, plug your device into your computer to access eBooks that were not purchased from the NOOK Store, manage an address book, and connect to several social networks, like Facebook and Gmail. There's also Barnes & Noble's LendMe Program, which we talked about in Chapter 24, *Getting Social*. You can also hide books from the LendMe list, which we'll talk about here.

We're also going to get really technical in this chapter. Here's where you'll find out how to use NOOK Color outside of the United States, and how to update the entire device with the latest software.

Taking Advantage of Barnes & Noble Stores

There are unique benefits to using your NOOK in Barnes & Noble stores.

Bring your NOOK Color into a Barnes & Noble store, and you can do several things:

- Use Wi-Fi for free.
- Read books for free for up to one hour per day.
- Get technical support for your NOOK Color.
- Return a device for repair.
- Download exclusive content.
- Get a personalized recommendation for a new book.
- Try out new NOOKs.

Perhaps the biggest benefit is being able to use Wi-Fi for free! Through a special arrangement with AT&T, all Barnes & Noble stores have free Wi-Fi. Let's say you don't have a local wireless connection at your home or at any nearby coffee shops. You can just come to your nearest Barnes & Noble and download whatever books, periodicals, or apps you've had your eye on.

You can also read books for free. As we talked about earlier in the book, you can always download samples of books, periodicals, or apps. However, if you bring your NOOK Color into your local Barnes & Noble, you can also download entire books for reading. Barnes & Noble gives you a full hour with a book of your choosing. Your NOOK Color will let you know when the time is up.

> Barnes & Noble gives you an hour a day per book to read for free, so you can always come back the next day to read more!

Barnes & Noble booksellers are known for being voracious readers as well as hearty app users, so visiting a local store is an excellent way to find out about the latest and best content for your NOOK. Not sure if the newest book from your favorite novelist is worth reading? Trying to decide on the best international newspaper to read for an upcoming trip? Come into the store and chat with a Barnes & Noble bookseller; they'll help get you going in the right direction. And while this book hopes to answer all your NOOK questions, sometimes it helps to have someone give you hands-on assistance. The folks at Barnes & Noble stores know all about your device and are happy to help answer your technical questions.

Barnes & Noble stores also offer a faster way to get your device repaired. If you chat with a Barnes & Noble bookseller and still aren't able to get your NOOK Color up and running, the bookseller may suggest sending the device in for repair. Before you do, you'll want to ask the bookseller:

- What the problem seems to be
- If the problem falls under warranty

You might also want to read over the FAQ as well as the warranty information (basic and extended) that are listed at the very end of this book. If the problem doesn't fall under warranty or if the warranty has expired, Barnes & Noble won't be able to repair your device. It's important that you know these answers before you hand it off to the bookseller to ship in for repair. However, if your warranty is current and the problem falls under the warranty, the assistant will be happy to send your NOOK Color in for repair.

Come into a Barnes & Noble store and you can also download exclusive content. There are thousands of books, periodicals, and apps in the NOOK Store, and dozens more come in every day. But Barnes & Noble also offers interesting, exclusive goodies that can only be downloaded at the store. No need to plug in your NOOK Color—as long as you are in the store, the local Wi-Fi will give you access to the content. You can ask the Barnes & Noble booksellers what exclusive content is available at the moment.

You can also take the new NOOKs for a test drive. If you have a NOOK Color, why not check out the lower-priced NOOK Simple Touch or the powerful, versatile NOOK Tablet? It's a fun way to learn all about the latest and greatest capabilities offered by eReading.

Background Images

Tired of the picture in the background? You can switch up the wallpaper anytime. Go to the Home Screen and hold your finger on any empty part of the screen—in other words, directly on the wallpaper. You'll get the option to Change Wallpaper.

There are three ways to change the background:

- Wallpaper
- Photo Gallery
- Live Wallpapers

Wallpapers are static pictures that sit in the background as you use your NOOK Color. Your device comes stocked with a handful of wallpapers that you can choose from.

Photo Gallery has photos that you have downloaded, either from someone else or from your own personal collection. Your NOOK Color also comes with a few sample pictures.

Finally, Live Wallpapers are cool animated pictures that move in the background as you use your NOOK Color. These are available from retailers in the NOOK Store.

If you'd like to upload your own photos, plug your NOOK Color into a computer, and wait for the machine to read the device. Once you're connected, go to the file marked MyComputer and double click on the NOOK folder. Now you are looking into your NOOK Color memory.

Depending on what you want to add, click on the file marked Wallpapers or Pictures. Here, create a new folder with a one-word file name. Double click on it, and add any image files you want. This is best done by right-clicking Copy on the image, and by right-clicking Paste in the file you've created.

Image files the NOOK Color supports:

▶ JPEG

▶ GIF

▶ PNG

▶ BMP

Reading PDFs and Other Documents

NOOK Color isn't restricted to only reading eBooks and other eReader files. The device can also read PDF files. The process of adding them onto NOOK Color is easy, and very similar to adding ePub files (as detailed in Chapter 10, *Memory and Storage*). In order to load PDF files to your NOOK, start by plugging your NOOK Color into a computer.

Next, right-click and copy the PDF file you want to add, and paste it into the NOOK file labeled MyFiles. After that, safely disconnect NOOK Color, push the Quick Nav button, and tap Library. From there, navigate through the folder tab on the top left corner, locate and tap MyFiles. You should then be able to locate all of the ePub files that weren't downloaded through the NOOK Store. The PDF files are marked with an icon featuring a PDF logo.

Once you've found your PDF file, double click on it, and it should open up just like an eBook. (Note that you can't search for text in PDF files saved as images.) In fact, the layout is exactly the same as an eBook. (Note that you can't search for text in PDF files saved as images.) Simply read the PDF file the same way you read an eBook, and there should be no problems. If there are, however, consult Chapter 19, *Troubleshooting*.

Updating Your NOOK Color Software

Keeping your NOOK Color updated with the latest software is both simple and important to do. It allows you to benefit from any changes Barnes & Noble has made to make NOOK Color a better product, and to keep everything running smoothly.

NOOK Color actually updates on its own. When you're connected to Wi-Fi, it will automatically download any updates without any input from you. When an update has been installed, a new button will appear on the lower left of the screen, right next to the button that takes you back to your most recent book. Tapping the button will let you know that a new version of NOOK Color software was successfully installed.

If you want to know what software version your NOOK Color is using—this is important sometimes for troubleshooting and discussing new features that may be added to NOOK Color—the process is simple. Push the NOOK button, tap Settings, then tap Device Info. From there, tap the button labeled About Your NOOK. The software version should be listed in the middle of the table, right under your account address and the model number.

27.
Troubleshooting

NOOK Color is a user-friendly device, but it can occasionally run into some technical problems. The following are a few fixes that should help you solve potential problems.

Not Charging?

If NOOK Color isn't charging, make sure the device is plugged in. You'll know it is being charged when the orange light at the bottom of the device is lit up. When the device is fully charged, the light turns green.

Not Connecting to Wi-Fi?

If you're having trouble connecting to Wi-Fi, make sure that your NOOK Color is looking for the desired Wi-Fi. Push the NOOK button, press Settings and tap Wireless. From there, make sure the Wi-Fi is turned on and is connected. If not, either connect to an alternative Wi-Fi network, or troubleshoot your wireless router.

Music or Video Isn't Playing?

Your NOOK Color is a versatile device, but it can only understand certain music and video files. As far as music, it can play AAC, amr, mid, MIDI, MP3, M4a, wav, and Ogg Vorbis files. With video, it plays Adobe Flash, 3gp, 3g2, mkv, mp4, m4v, MPEG-4, H.263, and H.264 formats. Other video types aren't guaranteed to work.

What Kind of Customer Support Does Barnes & Noble Offer?

Barnes & Noble understands that NOOK Color is an investment on your part, so the company has provided several ways for you to get the technical support you need:

- Phone
- Email
- Live online chat
- In-store
- NOOK Web site FAQs

If you'd like to talk with a Barnes & Noble representative over your landline or cell phone, give the company a call at 1-800-843-2665. If you prefer email, the Barnes & Noble troubleshooting email is nook@ barnesandnoble.com.

You can also talk with someone online, which is like using an instant messenger program to ask an expert your pressing questions. To use the live online chat, go to http://www. barnesandnoble.com/nook/ support/ and click on the Chat Now link under Chat With a NOOK Color expert. The Web site will ask you for five pieces of information:

- Name
- Email address
- Product (NOOK Simple Touch, NOOK Color, or NOOK Tablet)
- Serial or order number
- Question

Give your full name and the email address that is associated with your Barnes & Noble account. Be sure to choose the right product, as the troubleshooting for your NOOK, NOOK Tablet, and NOOK Color are different.

Also, it helps if you have your serial number. If your NOOK Color is functional, tap the NOOK button to bring up the Quick Nav Bar, and then tap the Settings icon. Next, tap the Settings icon, touch

Device Info, and then About Your NOOK. Here you'll find lots of details about your device. The serial number will be the second-to-last number listed. It will be 16 digits.

Finally, describe your problem as thoroughly as possible. You'll be talking to someone—that's the whole purpose, of course!—but asking a clear, detailed question will make the process faster and easier. The more info the customer service representative has, the better they can help.

When you're ready, tap the Submit button. A Barnes & Noble representative will then hop online and answer your question.

And, if you'd like to talk to someone face to face, you can go online at barnesandnoble.com and find a nearby Barnes & Noble store. The knowledgable associates will be happy to help.

Other Issues

If your problem is not pressing, you can also use the Barnes & Noble forums. In fact, there's always an active community on the forums, so your question may have already been asked by someone else and answered by Barnes & Noble. You can check out the troubleshooting forums at: http://bookclubs.barnesandnoble.com.

28. Accessories

With all that you'll be using your NOOK Color for, you'll want to make sure that it looks great (and is protected) while you're carrying it around town with you. Barnes & Noble offers many options to protect and accessorize your NOOK Color, including cases, bags, and covers that will help your NOOK Color stand out.

Covers

Barnes & Noble sells an antiglare cover that can be simply applied on top of NOOK Color. The antiglare cover helps protect the screen from dust and foreign particles while the screen maintains all of its touch functionality.

NOOK Color also can be put into a selection of cases. These simply slide over the device to add some color, protection, and style. Some can open from left to right like a book, while others don't have a spine at all. They come in leather and cloth and include designs from famous designers like Jonathan Adler and Kate Spade.

Bags

In addition to protective cases, Barnes & Noble also sells entire bags with pockets made for holding NOOK Color. These cloth bags will keep NOOK Color in a safe place, and also have room for laptops up to 15 inches.

Other

Barnes & Noble also sells a stand for NOOK Color. This lets readers use the device and keep it at an optimal reading angle without needing to hold it.

Barnes & Noble sells car chargers for your NOOK Color, as well as extra AC adapters, to ensure the device can get power at any and all times.

NOOK Simple Touch

29. A Look at NOOK Simple Touch

NOOK Simple Touch (referred to as simply "NOOK" throughout this section) is designed purely for reading. The black-and-white touchscreen uses an E Ink display, a popular process that makes reading less strenuous on the eyes. If you often find yourself reading for hours at a time, the NOOK may be the best device for you.

The dimensions make almost a perfect square six inches, and the thin frame, under a half-inch thick, makes it easy to throw NOOK in your bag and take it wherever you go.

The front of NOOK is taken up by the black-and-white touchscreen. Right below the touchscreen is the "n" symbol. The "n" key is the Quick Nav button. Tap it when the power is on and your current options will pop up onscreen. Don't worry about getting lost in the menus as the Quick Nav menu is always a touch away.

Check out the buttons on either side of the touchscreen. There are two on the left and two on the right. These are the Page Turn buttons. The buttons are indicated by grooves. In NOOK's default setting, the top two buttons turn the page forward, and the bottom two turn the page back. They also scroll up and down on the Library screen. Whether a particular button scrolls up or down, however, can be changed in Settings, which we'll discuss later in this section.

Now tilt NOOK away from you so you can see the thin bottom right below the the Quick Nav button. Here is the hole for your USB cable. We'll plug it in after we finish our tour of the device.

Turn the device to the right and again to the right. You'll be looking at the top of NOOK. Here you'll find the Power button. You don't have to turn the power off whenever you stop using it— it will automatically power down after you leave it idle—but you do need to use the Power button to turn it on! We'll turn your NOOK on soon.

Flip NOOK so the touchscreen is facing down. Check out the textured back. While other devices are smooth and, arguably, slippery, NOOK's raised back makes it easier to keep a grip while you read your favorite book, magazine, newspaper, or document.

Finally, let's take a look at the microSD card slot. Flip your NOOK right-side-up and look at the right side of the device. You should see removable rubber cover, under which is a compartment for the microSD card.

You'll be downloading lots of books and periodicals for your NOOK, and they can take up a lot of memory. The microSD card allows you to expand the memory your NOOK can handle. We'll discuss microSD cards and download management further in Chapter 32, *Memory and Storage*.

In the Box

Inside the NOOK box you'll find everything you need to get started. The start-up process will only take you a few minutes and, aside from optional accessories, you don't need to make any other purchases. We'll look at NOOK covers and other fun accessories later in the book, but for now let's crack open the box and see what's packed inside. First, take off the wrapper and, then, bend the bottom half of the box to open the package. Here's what you'll find:

▶ The NOOK device

▶ A custom USB-to-micro-USB cord

▶ A wall plug

Don't see the cord and the wall plug? NOOK is at the top of the box, while the cord and the wall plug are snug inside the bottom of the container. (Depending on when you bought your NOOK, the wall plug may not be included; it can be bought seperately.)

Before You Start: Charge It

Okay, now the tour is done, and you'd probably like to start reading! It's hard to resist the urge to dive right in when you get a new device, but there is usually a step or two you need to take to guarantee a good experience. For NOOK, Barnes & Noble recommends that you charge up your device before you start playing with it.

Let's get it charged before we take it for a spin. Take a look at the USB cord. One end has a traditional USB format that connects to most computers, while the other end has a micro-USB connection that is commonly used by devices. Plug the wider USB end into NOOK's wall plug.

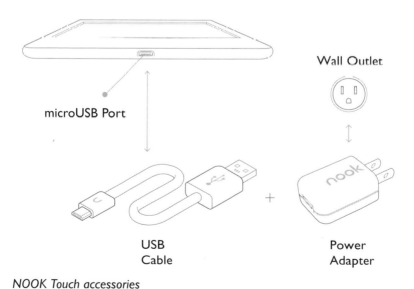

microUSB Port

Wall Outlet

USB
Cable

Power
Adapter

NOOK Touch accessories

On the other end, the micro-USB end plugs into your NOOK. As you may remember, look at the "n" key and you'll find the micro-USB hole right below it. Connect the micro-USB in there and, finally, put your wall plug into an outlet. If everything is connected properly, you'll see a cool orange glow from a tiny light next to the microUSB connector on your NOOK. The glow will turn green once NOOK is fully charged.

Powering Up

If the light on the bottom of NOOK changed from orange to green, your NOOK is charged. You can finally turn it on! Press and hold the Power key, which is located on the back side of the device for two seconds. You should get a brief loading screen and then the Intro page.

Getting started requires four steps:

▶ Agree to Barnes & Noble Terms and Conditions.

▶ Set time zone.

▶ Add Wi-Fi.

▶ Register your NOOK and default credit card.

Let's go through the Intro steps. The Barnes & Noble Terms and Conditions is the list of things you promise to do (and not do!) with the device. It may seem like a long agreement, but at least skim through the document so you can feel 100 percent comfortable using the device. When you've looked it over, tap on the Agree icon.

Wi-Fi

The next big step is setting up your Wi-Fi connection.

A Wi-Fi connection is required to finish setting up your NOOK. Are

NOOK Home Screen

you in a Wi-Fi hotspot? A hotspot is any area where a Wi-Fi router is in range. Wi-Fi routers at public venues are usually free to join, but others, like the Wi-Fi router in your house, usually are (and should be) password-protected. It's unlawful to use someone's personal Wi-Fi without his or her permission.

The good news is that every Barnes & Noble store offers free Wi-Fi. It is a great alternative if you don't have your own Wi-Fi router at home. When you are in a hotspot, NOOK will ask you to choose a Wi-Fi connection. If you are at a Barnes & Noble store, a bookseller can help choose the best one—in fact, that's true of any public venue. If you are at your home hotspot, choose your Wi-Fi connection name from the list. If the Wi-Fi router requires a password, NOOK will ask you to type it in. A virtual keyboard will pop up onscreen that will let you punch in the letters and numbers in the password.

If you are having trouble with your home Wi-Fi showing up on the list of hotspots, try unplugging your router from the outlet and plugging it back in after five minutes. Check the list of NOOK hotspots again and it should be on the list.

Account Information

The last step is registering your device, your account, and your credit card. Do you already have a barnesandnoble.com account? If you remember your username and password, type them in here. If you have an account, but don't remember the password, you can hop online and request a new password to be emailed to you.

If you don't have an account yet, go ahead and type in a username and a password that you will remember, but others will have a hard time guessing. Barnes & Noble will check to see if anyone else is using the same username. As soon as you have an original username, you'll move on to the credit card.

NOOK wireless setup

Like other mobile devices, your NOOK allows you to make purchases without having to type in your credit card information each time. Instead, you type in a default credit card number that will be stored privately by Barnes & Noble. Anytime you make a purchase, the amount will be charged to your card. Keep in mind that you can always change your default card or even input another card for a particular purchase—it's just much easier to have a credit card on file. Choose a valid credit card, type in the account number, and, once it is validated with a quick online check, you're ready to begin using your NOOK!

Navigating the Touchscreen

Your NOOK can understand four different types of gestures on its screen. If you move your finger up or down, the NOOK will scroll through its menus. If you swipe your finger left or right, the pages will turn forward or back.

You can also tap the screen to access a book, or to open up the book's menu. The menu will let you change the size of the font, find a specific word, or access the table of contents. You can also quickly tap the screen twice to receive more information about the books at the NOOK Store.

The last gesture you'll need to use on NOOK's touchscreen is the Press and Hold gesture. If you Press and Hold a word while reading, you can highlight phrases, add annotations, or look up the word in NOOK's built-in dictionary.

Quick Nav Button

NOOK's Quick Nav button, the "n" key at the bottom of the device, opens up the main menu, also known as the Quick Nav Bar.

These Quick Nav Bar icons are the easiest way to navigate through the most commonly used features of your NOOK. Tap the Quick Nav button and the following icons appear on the bottom of the screen:

- **Home:** Access to NOOK's home page
- **Library:** Quick access to the books and samples on NOOK
- **Shop:** Takes you to the NOOK Store to purchase more reading
- **Search:** The Search Bar lets you look for items in both your NOOK Library and the NOOK Store.
- **Settings:** Accesses the Settings page to change several NOOK features.

NOOK Quick Nav Bar

The Status Bar

Like most devices, your NOOK includes an active area that provides helpful info like how much battery life is remaining, what you were last reading, and so on. This is the Status Bar.

Located at the top, the Status Bar shows you the following:

- ▶ The bookmark: A placeholder you left at a particular place in a book, magazine, or newspaper. Tap it and it will open the reading material to where you marked it.
- ▶ Wi-Fi, power, and clock.

NOOK Status Bar

The first icon, which looks like a fan, is your Wi-Fi icon. When the Wi-Fi is active, you'll see little waves coming from the bottom to the top of the icon; if you have any other Wi-Fi-enabled devices, you are probably familiar with this symbol. If the fan is empty, however, then you don't have Wi-Fi right now. Remember that it doesn't mean that Wi-Fi isn't enabled, but that you aren't getting Wi-Fi right now. Your local hotspot could be down or there could be some other technical difficulty. If you're having trouble, try going through the previous Wi-Fi process again or skip ahead to Chapter 34, Troubleshooting, to get yourself squared away before proceeding.

The second icon, which looks like a little battery, is your power icon. If you were able to get your NOOK fully charged, the battery will look totally full. As the device is used, the battery will begin to drain. No worries: Your NOOK can handle several hours of continuous use and can sit on standby for weeks. However, when your power finally does get low, the battery meter will shrink and, eventually, flash to warn you.

The third and final icon, the clock, gives you the current time.

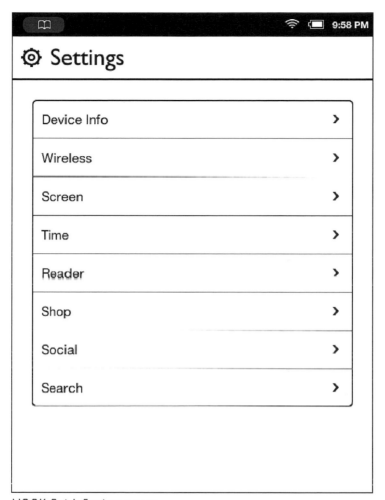

NOOK Quick Settings screen

Settings

To get to the Settings screen, just tap the Settings icon on the Quick Nav Bar. You will see the following list of settings:

- Device Info
- Wireless
- Screen
- Time
- Reader
- Shop
- Social
- Search

Here you can tweak and adjust your NOOK to truly make it your own. The Settings page affects nearly every aspect of your NOOK, from your reading experience to your ability to connect with other NOOK readers, so we'll be referring to this screen throughout the book. At this point, just remember that you can access it via the Quick Nav Bar under Settings.

Warranty Options

Finally, before you start having fun with your NOOK, you will want to consider the warranty options; Barnes & Noble has a couple different ones to choose from:

- Standard Warranty
- B&N Protection Plan

You automatically got the basic protection as soon as you bought your NOOK. It's up to you to decide whether you'd like to upgrade to the advanced protection plan.

NOOK Color Protection Plan
(based on http://www.nook.com/warranty)

	Standard Warranty (1 year)	B&N Protection Plan (2 years)
Customer Service	x	x
Rapid Replacement	x	x
Accidental Damage		x
Extended Service		x

Both plans include:

- ▶ Free customer support
- ▶ Rapid replacement if NOOK malfunctions
- ▶ Minimum of one year protection

However, the B&N Protection Plan adds:

- ▶ Two years of protection
- ▶ Rapid replacement for accidents like spills and cracks

Visit http://www.nook.com/warranty or call 1-800-843-2665 for the latest B&N Protection Plan pricing. For more details on the support plans, read the FAQ at the end of this book.

Now you have everything you need to get started with your NOOK. But what about reading books, downloading items, and managing your ever-growing collection? The rest of this section of the book is dedicated to making sure you get the most out of your device. Let's get started.

30. NOOK Store

There are a few books—such as Bram Stoker's *Dracula*—included on your NOOK from the get-go, but you probably want to read other books and expand your digital library. This is where the NOOK Store comes into play.

The NOOK Store is where you will be buying, sampling, and browsing through the NOOK Store's vast selection. The shop features all of the books and categories of books and periodicals you'd find at a Barnes & Noble retail store. After all, what good is an eReader with nothing to read? To access the NOOK Store, simply push the NOOK's Quick Nav button and touch the icon labeled Shop.

Be aware that while there are many free books in the NOOK Store (over a million free titles), most cost money. Buying them doesn't require you to re-input your credit card number, making new purchases quick and easy. You can set your NOOK to require a password for every purchase by going to the Settings menu and tapping the Shop menu.

The **NOOK** Store's Front Window

Three different sections appear in the NOOK Store's Front Window. The first one, Browse Shop, lets you peruse books, magazines, and newspapers by genre and category. It also lets you look through more specific search lists like the B&N Top 100, bestsellers, and even recommended titles based on your previous purchases.

The Popular Lists pane covers similar ground. This section reveals more details about the current major titles in the literary world,

NOOK Store

letting you know about new releases, and which books are on the *New York Times* best sellers list. It also lists featured books on sale. The final pane—the one on the bottom half of the screen—is a rotating bulletin board of information. This section details what's new, and other NOOK-specific information. Aside from B&N recommendations, this page might suggest fun summer reads, or offer monthly spotlights or discounts at Barnes & Noble retail stores. This is also the pane that updates the most often.

Browsing

Browsing through the NOOK Store is easy. By using lists, or simply by searching, countless books are waiting to be discovered. If you aren't sure what books to look for, it's best either to start with a list. Just tap the list/section you want and start narrowing your search. If you don't have a specific title in mind, we'll tell you how to find a book through the first pane.

Books

If you already know what book you're looking for, then use the Search Bar at the top right corner of the screen. Type in either the name of the book or the author's name, and you'll find a list of works containing the words you typed into the Search Bar. This function works like any other search engine. If you're currently browsing for something new and exciting to read, however, we recommend you go through the Store's detailed list to find a title that piques your interest.

Select the Book section from the Store's Front Window, and you'll find four pages of categories. Each category covers a different literary genre. From biographies to humor, every style is represented. Peruse each page until you see a genre that appeals to you and select it. You will be taken to another series of pages that break the genre down further. For example, select Humor and a list displaying the types of humor books, such as essays or cartoons. When you find the specific genre that sounds most interesting, press it, and individual books will be displayed.

Remember, samples are free. If you aren't sure about buying a certain book, it's worth checking out the sample to see if it's interesting.

Look through the pages until you see a title you think you might like. Tap the book's cover, it, and a brief summary of the book and publication info will appear. If you want to buy it, push the button that says Purchase. Then, the Purchase button will change to say Confirm. Press it, and the book will automatically be downloaded to your NOOK. If you aren't sure about buying the book, feel free to download the sample by pushing the button marked Sample.

After you've made a decision about whether or not to purchase the book or download the sample, simply press the Back button—the little button with an arrow pointing left—and look for other books to check out.

If you like the sample of what you've just read, you can get back to the purchase screen in the Store by double tapping the book in your Library. That way, you can easily buy the book without having to navigate through the search process from the beginning.

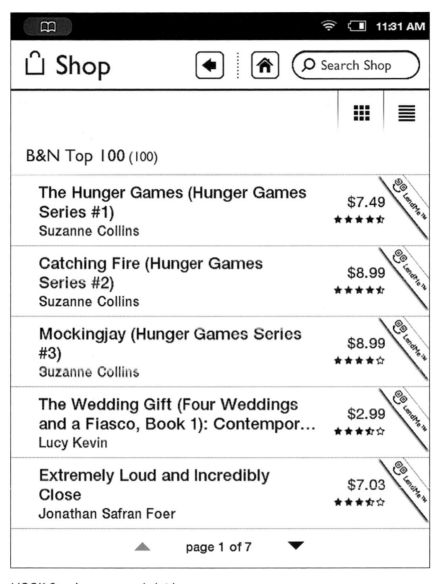

NOOK Store's recommended titles

Periodicals

You can browse and shop for magazines and newspapers in a way similar to books. The difference, however, lies in purchasing them. Instead of being able to download a sample, all publications will let you try a 14-day free trial, or you can just purchase the current issue. Once you're certain you want the magazine or newspaper, you can subscribe to it for a subscription fee, and each new issue will automatically download to your NOOK as soon as it is released. There is usually a steep discount for subscribing compared to buying each individual issue, and there's no stress about ever forgetting to download one. It guarantees a continuous supply of engaging new content for your NOOK.

In our next chapter, we'll talk about how to access and read all the books, magazines, and periodicals you've just sampled and purchased.

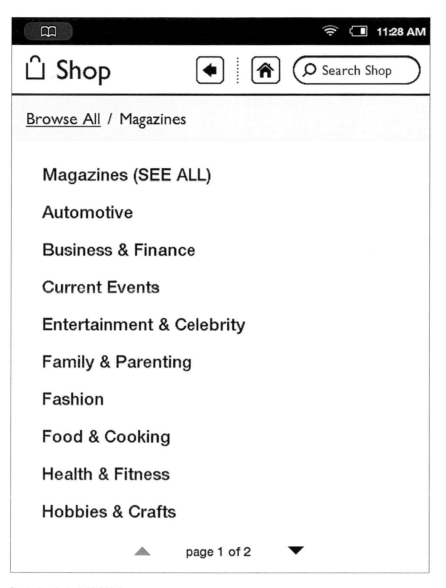

Periodicals in NOOK Store

31. Books and Periodicals

Now that you've purchased some exciting new titles and are on your way to building an impressive library on your NOOK, this chapter will show you how to dive into these great new reads and enjoy special reading features like bookmarking, looking up words, and more. We'll also discuss how to navigate and organize your own personal bookshelves and making books, magazines, and newspapers quick and easy to find.

Finding Books and Navigating the Library

NOOK Library can be accessed by pressing the Quick Nav arrow and tapping the icon marked Library. When you first tap the Library icon, a list will appear with all the reading you have downloaded to your NOOK. This list can be managed with the two navigation menus, found immediately under the Library title. The first menu helps navigate through books, periodicals and any files you have transferred from your personal computer; we'll explain this further in Chapter 32, *Memory and Storage*. The second menu reorganizes the list by title, author or the most recently added books in the list.

If you don't like the default layout of the Library, you can change it to look like a simple list of books. Near the top right corner of your NOOK Library screen, there are two buttons. One looks like nine squares arranged like a tic-tac-toe board and the other looks like a series of horizontal lines. Pushing either of them changes the appearance of the Library, so choose whichever one suits your preference.

If you're looking for a book you know you have, but isn't in your Library, be sure to check out your archived books. Archived books are not on your NOOK, but Barnes & Noble recognizes that you own them. The Archive tab can be found in the first navigation menu of the NOOK Library. Tap the button marked Archived and tap the Unarchive button to re-download items. To learn more about this, check out the section in Chapter 32, *Memory and Storage*.

Reading Books

To open a book, tap on its cover or its title in your Library. Now that a book is open, diving in to start reading is as easy as a tap or a swipe. Let's start by turning pages. The first way to turn a page on your NOOK is to use the buttons on either side of the device. The top two turn the page forward, while the bottom two buttons turn it back. You can also turn a page forward by swiping your finger from right to left, from left to right to turn the page back, or by tapping the right or left edges of the display.

Once you master page turning, you'll be able to navigate through anything NOOK has to offer. There are, however, many other functions for readers that are worth discussing. NOOK lets readers annotate their books, and offers several ways to expedite page-turning searches for specific parts of a book.

The side button configuration is the default for the device, but you can change it so that the top two buttons turn the page back and the bottom two buttons turn the page forward by pushing the Quick Nav button, tapping Settings, and then tapping Reader. The only option on this page changes the button configuration. However, for the rest of the book, we'll assume NOOK is using its factory defaults.

Notes and Highlights

Sometimes, part of a book is so impactful, you'll want to remember exactly where you read it, and that may not sound like the easiest thing to do on an eReader. After all, you can't fold down a page corner on your NOOK. Fortunately, NOOK has several key features that far surpass simple page-folding.

Notes and Highlights serve a similar purpose, except the former lets you add a few quick comments. Essentially, both features let you mark a word or passage, allowing you to access it from the Content page. This makes looking for specific parts of a book extraordinarily easy.

In order to drag a highlight through a series of words, hold your finger down on the first word, and drag it across until you're over the last word you want highlighted. The NOOK is a little bit finicky in regards to recognizing which word your finger is on, but after a few attempts, you'll get the hang of it.

Hold your finger down on a word for a second in a NOOK Book, and a submenu will appear at the bottom of the screen that lets you do several things to that specific word. You can highlight the word, add an annotation to the word, share the word with NOOK Friends, post it on Facebook or Twitter, or you can look the word up in NOOK's built-in dictionary.

Highlighting the word marks the word in a bold highlight. This helps mark key points of the book that you may want to return to later. Notes lets you add any comments you'd like to remember while reading the book. If there is a Note on a page, it will be marked with a little sticky note-like icon on the right side of the page. Tapping it will display the note you left.

Bookmarks

Bookmarks, like in real books, mark an entire page for later reading. NOOK Bookmarks don't single out a noteworthy word or phrase, or allow annotations. Instead, they simply mark a page so that you can return to it later.

When you tap a page in a NOOK Book, a small ribbon-looking tab with a lower case "n" appears in the top right corner. Tapping it highlights the Bookmark, and this page number will be saved for future reading. To undo a Bookmark, simply tap it a second time, and the ribbon icon should no longer be highlighted.

To access the page later on, go to the Content menu. The Bookmarks tab is the one closest to the right. Tap it, and a list of Bookmarks will appear on your screen.

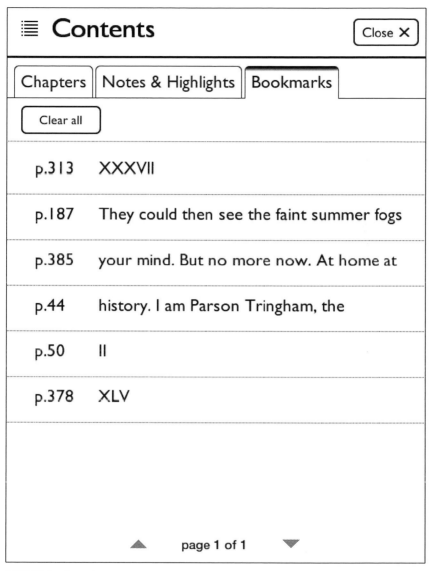

NOOK Bookmarks

Navigating **NOOK** Books

NOOK Books are extraordinarily easy to navigate. In addition to the previously mentioned Notes and Highlight, NOOK has several features to make finding the exact part of a book as simple as a tap. In order to start, tap the center of the page, and you'll find the following options in the Reading Tools menu:

> ▶ Content

> ▶ Find

> ▶ Go To

> ▶ Text

> ▶ More

Even with all of the in-book options, you can always push the Quick Nav button and navigate NOOK like you normally would. This will take you out of the book and into other pages like Settings, the Store, or even back to the Home Screen.

Content

The Content button brings you back to the table of contents. Here, you can easily jump to each chapter in the book, and navigate through the list in the first pane to see what each chapter is called and on what page the chapter begins. Tapping the chapter name takes you to that specific chapter.

The Content screen also has tabs in the upper section of the page to help navigate through your Chapters, Notes and Highlights, and Bookmarks. Find the other tabs near the top of the screen and tap them. The lists will look similar to the Chapter pane, but will feature

I

ON AN EVENING IN the latter part of May a middle-aged man was walking homeward from Shaston to the village of Marlott, in the adjoining Vale of Blakemore or Blackmoor. The pair of legs that carried him were rickety, and there was a bias in his gait which inclined him somewhat to the left of a straight line. He occasionally gave a smart nod, as if in confirmation of some opinion, though he was not thinking of anything in particular. An empty egg-basket was slung upon his arm, the nap of his hat was ruffled, a patch being quite worn away at its brim where his thumb came in taking it off. Presently he was met by an elderly parson astride on a gray mare, who, as he rode, hummed a wandering tune.

"Good night t'ee," said the man with the basket.

content	find	go to	text	more

NOOK Reading Tools menu

all the Notes, Highlights and Bookmarks you've already made in the book. These are great ways to personalize the navigation experience.

Find

The Find feature lets you look for any word in the book. Just type in the word and within seconds, you'll see a chronological list of places where the word appears in the book. The list displays part of the sentence in the which the word appears, as well as the page number.

To use the Find feature, open up the Reading Tools and tap the button marked Find. This will bring up a Search Bar. Type in the word you want to look for, and a menu similar to the ones found in the Content feature will appear, listing every instance of the word you just typed in. From here, tap the section of the book you want to go to, and NOOK will take you there.

> These features don't only work for NOOK Books. The Reading Tools menu works exactly the same way in PDF files and periodicals. (However, if a PDF has been saved as an image, its text is not searchable.)

Go To

The Go To feature adds a small slide bar at the bottom of the NOOK screen. This bar lets you navigate quickly to a specific page by simply moving your finger. Slide your finger left or right along the bar, and the book will either advance forward or back several pages, depending on how far you moved your finger. Once you let go, NOOK will tell you some brief details about the page, usually how many pages are left in that chapter.

This feature makes skimming through books remarkably fast. There are also two buttons in the bottom corners of the screen that let you either go back to the previous page, without losing your initial place in the book (Go Back), or pull up a keyboard into which you can type the number of the page you want to visit (Go to Page).

Text

There are five different options in the Text Settings menu. The first one, located in the middle of the screen, changes the font size—i.e., how large or small the words are on the pages. The next area—located at the bottom-left of the screen—changes the style of the font. Fonts include Trebuchet, Gill Sans, and Amasis. The two key options to the right change the line spacing and the margins of the text on each page. They change how closely spaced the lines of text are to each other and how close they are to the edge of the screen.

The final option, the little check-mark box on the bottom right, lets you set all of these options to the publisher's defaults. In most cases, the publisher's defaults set everything the way the book-publishing company wants it, setting the font, margins, and line spacing the way the publisher intended.

More

The More section lists additional details about the book, including the book's full title and the author. Beyond this, there are two distinct differences in regards to the way this page may look, depending on whether or not the book was downloaded from the NOOK Store.

If the book was downloaded from the NOOK Store, the page should emulate the book's page in the Store. As such, you can also use this page to share books, read reviews, and find similar books. If the book

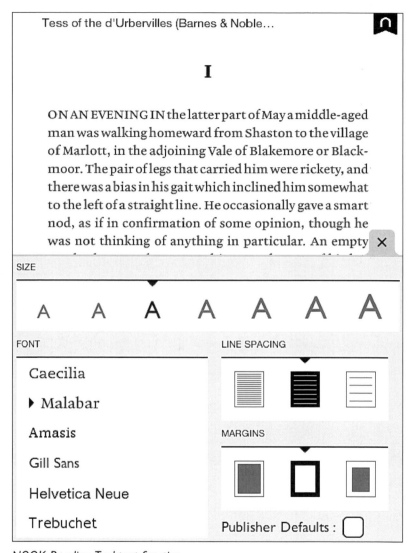

NOOK Reading Tool text function

wasn't downloaded from the NOOK Store, the page will display where the book is stored in NOOK's memory, and how large the file is.

Adding books from sources outside of the NOOK Store is a great way to increase your NOOK Library. There are many available public-domain books (books that are no longer copyrighted) or independent works by authors who are trying to break into the literary world; some of these works may not be available through Barnes & Noble. We'll talk about getting books from sources other than the NOOK Store later in this book, and address how to add them to your NOOK in Chapter 32, *Memory and Storage*. We'll also talk about where to download free books in the Appendix, under Online Resources.

Creating Bookshelves

To best organize and manage your NOOK Library, you'll probably want to create your own bookshelves. This way, you can personally manage the books any way you want. To do so, start by going to the View menu at the top of the Library screen, and tapping Shelves.

From here, tap the button near the top right corner marked Add Shelf. This opens a page that will let you name the new Shelf. We recommend a name that will be easy to remember and understand later, like a Shelf for each literary genre or the month you purchased the books. Type in the Shelf's name, and tap Save in the bottom right corner.

After you name your Shelf, a list will appear of all the books currently on your NOOK. Go through this list and check each book you want to place on that shelf by tapping the little square next to the book's title. A check mark will appear in the box if you

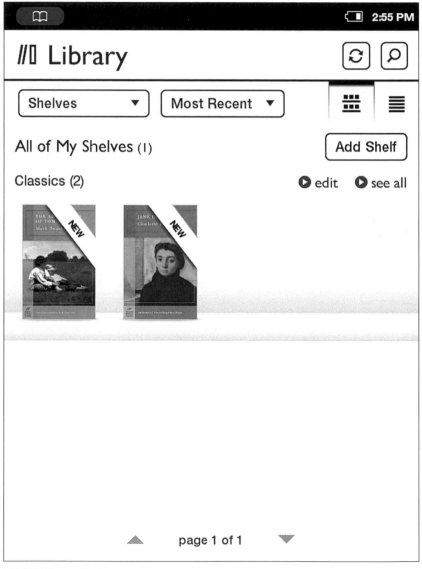

NOOK Library Shelves

tapped it correctly. Once you've checked each book you want on the shelf, tap Save, and the new shelf should appear.

If you ever want to add anything to the shelf, tap the Edit button right under the Add Shelf button. The Shelf page only shows four books on its front page, but you can tap the See All button to view all the books on the Shelf.

Although the process of opening PDF files works the same way it does with books, PDF files are marked with a large PDF logo, instead of featuring a book cover.

Borrowing Books

One of the biggest complaints about digital books is that, unlike traditional books, they can't be borrowed. And one of the biggest benefits of using the NOOK is that you can now borrow books!

There are two ways to borrow books:

- From a NOOK Friend
- From your local library

With NOOK Friends, you can borrow books from your connected friends and lend books to them, too. All it requires is sending a friend invitation to a real-life friend, via NOOK, to get connected. Your friend needs to:

- Be a registered user at the Barnes & Noble Web site
- Have a registered NOOK device or NOOK eReader app including NOOK for iPad, NOOK for PC, NOOK for Android, NOOK for iPhone, and NOOK for Mac.

To learn more about NOOK Friends and the LendMe program, check out Chapter 33, *Advanced Techniques*.

Periodicals

One of the best features of NOOK is its ability to download periodicals, giving you easy access to the latest newspaper and magazine issues. A new periodical you've subscribed to will download automatically when you're connected to Wi-Fi. The NOOK Store has plenty of magazines and newspapers from cities both big and small.

To purchase a periodical, navigate through the Store's menus—or simply search for a specific magazine or newspaper—and tap either Subscribe or Buy Current Issue. Subscriptions cost significantly less than the price of purchasing individual issues month after month. As discussed in the previous chapter, if you want to sample a magazine or newspaper before you subscribe, many periodicals offer a free 14-day trial subscription that will let you test-drive the publication before committing to a full-year subscription.

Reading Periodicals

By this point, you're probably used to reading books on your NOOK, and magazines aren't too different. After purchasing a magazine (or magazines), go to the Library and tap the issue you want to read. The magazine's cover will be the first thing to appear.
Periodicals on NOOK are designed for easy navigation. The biggest difference between periodicals and books is that newspapers have subsections, each of which has its own table of contents. After purchasing an issue of a newspaper, look through the table of contents for a section that seems interesting. Search through the

pages of stories until you find one that you would like to read, tap the headline, and you'll be presented with the article. Once you've finished reading the article (it may be several pages long), the bottom of the last page will show Next Article and Previous Article buttons. Tap them to access the next or previous pieces. Or, you can return to the table of contents by tapping anywhere on the screen to access the Quick Nav menu, then tap Content.

Archiving Periodicals

Periodicals are updated frequently, so it's easy to be overwhelmed by the amount of content that can clutter up your NOOK's Library. In order to avoid this, you can archive periodicals the exact same way you can archive a book. In your Library, double tap the periodical you want archived and you'll be brought to the Detail menu with the Archive button in the middle area of the screen. Press that, and the magazine or newspaper will be removed from your NOOK and placed in the B&N Archive until you want to re-download it. We'll cover re-downloading in Chapter 32, *Memory and Storage*.

Unsubscribe from a Periodical

Unfortunately, there isn't yet a way to unsubscribe from a periodical directly from your NOOK. The only way to do so is by going through Barnes & Noble's Web site. From the Web site, however, it is a very simple process to unsubscribe.

Simply go to BN.com and log into your account; click on Manage Subscriptions, and click the Cancel Subscriptions button for each periodical you want to stop receiving. You'll be able to keep whatever issues were downloaded during the time you were subscribed, including those downloaded during the free trial.

32. Memory and Storage

All digital gadgets have a limited amount of memory, and over time the memory gets used up. NOOK has a lot of space for books—typically about 2,000 of them—but the Library can get cluttered. It's worth knowing how much memory you have, and how you can help mitigate digital overflow.

In order to see how much memory is available for more books, magazines, or newspapers, access the Settings menu by pressing the Quick Nav button to open to the Quick Nav Bar, then tap Settings. Now tap the tab that says Device Info.

The Device Info pane shows how much battery power is left, and how much storage is left. NOOK's memory can hold a huge amount of books, but it's important to know how to manage and delete items just in case your NOOK begins to run out of storage space. In order to see how large a specific book is, simply double tap a book in the Library. The screen will tell you where it is stored and how large it is. The page will also give you information such as the title, author, and when the file was last modified.

Managing Space

Now that you know how much storage space is left, it's time to learn how to manage it. There are three places to store books:

▶ On your NOOK
▶ Online in the Barnes & Noble Archives
▶ On a microSD card

Archiving Items

To store a book in the Barnes & Noble Archives, go to the book's detail page by double tapping the book's cover in the Library. Then tap Archive. You should receive a confirmation message asking if you really want to do this. Tap Yes, and the book will be removed from your NOOK and stored on your account page on Barnes & Noble's Web site.

> You can only archive books that you've purchased from the Barnes & Noble store, including samples. Anything that you've put on your NOOK from your computer can be added and removed at your own discretion. We'll discuss how to connect your NOOK to your computer later in this chapter.

NOOK Device Info

The Barnes & Noble Archive

The Barnes & Noble Archive is a list of books that you own but aren't currently on your NOOK. To access the Archive, go to BN.com and log into your account. Once there, move your mouse over the My Account tab and click the link labeled My Library. On the left pane, there will be a button to view your Archive. In the Archive, you can delete anything you no longer want. The B&N Library also lists what samples you have downloaded, and what books have been purchased in full.

Re-downloading Items

Once your item is archived, it's extraordinarily easy to get a book back onto your NOOK. To do so, simply go to your NOOK Library, and—in the first scrolling menu—tap the tab labeled Archived. Tap what book you want to unarchive and the book starts to re-download onto your NOOK. It's a one-tap process and the book should be in your Library in a matter of seconds.

> Your NOOK contains both a Quick Start Guide and a User's Guide. These cannot be archived because they are there to answer questions you might have, as you are using your device.

MicroSD Cards

The final way to manage space on your NOOK is with microSD cards, which can be bought at most electronics stores and can be used to store an additional 32 GB of books and personal files on your NOOK. In order to use one, you first need to install an SD card. To do so, turn off your NOOK, lay it facedown, open the small compartment on the upper left side.

Now that the card is installed, you have access to any files stored on the card. In your Library, go to the My Files tab in the upper left corner, tap it, and look for the button labeled Memory Card. Here, you'll be able to find all the PDF and eReader files on the card that NOOK can read.

If you want to transfer files to your NOOK, you'll have to connect it to your computer.

In this section, we're talking exclusively about how to connect your NOOK to a Mac or PC. The process could be different for a Linux system.

Connecting NOOK to a Computer

Connecting NOOK to a computer is easy. Just take the USB-to-micro-USB cable that came with the device, plug the micro-USB end into your NOOK and the USB end into the computer. And with that, the two are connected. NOOK doesn't need any fancy installation, and will automatically work without any supporting programs.

From here, you can manually add files to your NOOK. To do so, simply right-click and select Copy for the files you want to put on your NOOK, navigate through the file named NOOK, and paste them in. It doesn't matter what subfiles they're in, as long as they're in a supported format and are stored on your NOOK. From there, the files should be accessible via your NOOK's library.

By connecting your NOOK to a computer, you'll be able to add free eBooks from Web sites, such as the Gutenberg Project, which we'll discuss later in the appendix.

eReader files the NOOK can read include:

- ePub
- PDF

JPEG, GIF, BMP, and PNG files are used for creating screensavers. To make a screensaver, create a folder with a one-word name in the MyFiles folder labeled Screensavers. Next, go to Settings through the Quick Nav menu, tap Screen, and you'll be able to select a new screensaver.

Wishlist

If you want to make a note to purchase something later, then Wishlists are the way to go. On the page for every book, right next to the book's star rating, there's a little box labeled My Wishlist. Tap the box to put a check mark in it. Now, every other book you've check marked will be on your Wishlist. The Wishlist is accessible from the front page of the Shop. With the Wishlist, you shouldn't have any trouble remembering which books to purchase.

Managing space is paramount once your NOOK starts overflowing. You'll be able to fit thousands of books in the allotted memory. Still, knowing how to sync your NOOK and archive books is key to keeping an organized Library.

33. Advanced Techniques

NOOK is filled with outstanding features and capabilities, many of which take you well beyond the simple enjoyment of reading. There are ways to create a personalized screensaver for your NOOK or plug the device into your computer to access eBooks that were not purchased from the NOOK Store.

There's also Barnes & Noble's LendMe Program—which we talked about in Chapter 31, *Books and Periodicals*. You can also hide titles from the LendMe list, which we'll talk about here.

We're going to get technical in this chapter. You'll learn how to use the NOOK outside of the United States and how to update the entire device with the latest software.

Taking Advantage of Barnes & Noble Stores

There are unique benefits to using your NOOK in a Barnes & Noble store. Bring your NOOK into a Barnes & Noble, and you can do several things:

- Use Wi-Fi for free.
- Read books for free for up to one hour a day.
- Get technical support for your NOOK.
- Return a device for repair.
- Get a personalized recommendation for a new book.
- Download exclusive content.
- Try out new NOOKs.

Perhaps the biggest benefit is being able to use Wi-Fi for free! Through a special arrangement with AT&T, all Barnes & Noble stores offer free Wi-Fi. Let's say you don't have a local wireless connection at your home or at any nearby coffee shops. You can just come to your nearest Barnes & Noble and download whatever books, magazines, or newspapers you've had your eye on.

You can also read books for free. As we talked about earlier in the book, as long as your NOOK is connected to Wi-Fi, you can always download samples of books or periodicals. It could be, say, the first 20 pages of a mystery novel or a 14-day trial magazine subscription. However, if you bring your NOOK into your local Barnes & Noble, you can download entire books for reading. Barnes & Noble gives you a full hour with a book of your choosing. Your NOOK will let you know when the time is up.

> Barnes & Noble gives you an hour per day per book to read for free, so you can always come back the next day to read more!

Barnes & Noble booksellers are known for being voracious readers, so visiting a local store is an excellent way to find out about the latest and best content for your NOOK. Not sure if the newest book from your favorite novelist is worth reading? Trying to decide on the best international newspaper to read for an upcoming trip? Come into the store and chat with a Barnes & Noble bookseller, they'll get you going in the right direction. And while this book in your hands attempts to answer all of your NOOK questions, sometimes it helps to have someone give you hands-on assistance. The folks at Barnes & Noble stores know all about your device and are happy to answer your technical questions.

Barnes & Noble booksellers can also help you get your device repaired if needed. If after discussing a problem with a bookseller, you still aren't able to get your NOOK up and running, the bookseller may suggest sending the device in for repair. Before you do, you'll want to find out:

- ▶ What the problem seems to be
- ▶ If the problem falls under warranty

It may help for you to read over the FAQ listed at the end of this book, as well as the warranty information (basic and extended) available on the Barnes & Noble Web site. If the problem doesn't fall under warranty, or if the warranty has expired, Barnes & Noble won't be able to repair your device. However, if your warranty is current and the problem falls under it, a bookseller will be happy to send your NOOK in for repair.

Come into a Barnes & Noble store and you can also download exclusive content. There are thousands of books and periodicals in the NOOK Store, with dozens more coming in every day. But Barnes & Noble stores also offer exclusive goodies that can only be downloaded at the store. No need to plug in your NOOK—as long as you are in the store, the Barnes & Noble Wi-Fi will make sure you have access to the content. You can ask the Barnes & Noble booksellers what exclusive content is available at the moment.

You can also take the new NOOKs for a test drive. If you have a NOOK, why not check out the powerful, versatile NOOK Tablet and NOOK Color? It's a fun way to learn about all of the latest and greatest capabilities offered by eReading.

Screensavers

The Screensaver is that series of images that appears when your NOOK is idle. If you don't touch the device for a while, or if you simply push the power button, the screensaver activates. There are two screensaver image series included with NOOK from the get-go: Authors and Nature. If you want, there is an easy way to change the screensaver to any series of images you'd like. Start by plugging your NOOK into a computer, and wait for the machine to recognize the device.

Once you're connected, go to the file marked My Computer and double click on the NOOK folder. Now you are looking into your NOOK's memory. Click on the file marked Screensavers. Here, create a new folder with a one-word file name. Double click it, and add any image files you want. This is best done by right-clicking Copy on the image, and by right-clicking Paste in the file you've created.

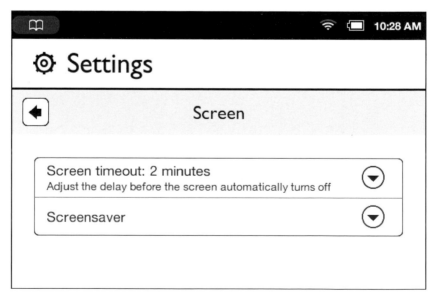

NOOK Screen settings

NOOK is a grayscale device. Any images you put in your screensaver will be displayed in black and white. Color images will still appear, but they may not have the same vibrancy they did on your computer.

Image files the NOOK supports include:

- JPEG
- GIF
- PNG
- BMP

Once you've added the images you want for your screensaver, remove your NOOK from the computer, push the Quick Nav button, tap Settings, then Screensavers. You'll see the Select menu for your screensavers. The file you created and put images into should be in that list.

Reading PDFs and Other Documents

NOOK isn't restricted to reading only eBooks and other eReader files. The device can also read PDF files. Adding PDFs onto your NOOK is easy. Start by plugging your NOOK into a computer. Next, right-click and copy the PDF file you want to add, and paste it into the appropriate subfolder in NOOK's MyFiles folder. After that, disconnect your NOOK, push the Quick Nav button, and tap Library. From there, navigate through the folder tab on the top left corner, locate and tap My Files. From there, you should be able to locate all of the ePub files that weren't downloaded through the NOOK Store, as well as PDF files. Once you've found your PDF file, tap on it, and it will open up just like an eBook.

Updating Your NOOK Software

Updating your NOOK is both simple and important to do. It allows you to benefit from any changes Barnes & Noble has made to make NOOK an even better product, and to keep everything running smoothly. In this section, we'll explain how to make sure your NOOK is updated.

NOOK actually updates on its own. When you're connected to Wi-Fi, it will automatically download any updates without requiring you to do anything special. When an update has been installed, a New button will appear on the top left of the screen, right next to the button that takes you back to your most recent book. Tapping the button will let you know that a new version of NOOK software was successfully installed.

If you want to know what software version your NOOK is using—this is important sometimes for troubleshooting or learning about new features added to your NOOK—the process is simple. Push the Quick Nav button, tap Settings, and then tap Device Info. From there, tap the button labeled About Your NOOK. The software version should be listed in the middle of the table, right under your account address and the model number.

Social Settings

NOOK, NOOK Color, and NOOK Tablet all work well as social devices. There are multiple features and ways to keep you connected to your friends, and to help encourage their reading habits. Barnes & Noble has included several services to help cultivate and encourage a literate community of NOOK users.

Ready to get social? In the Settings menu, tap Social, and you'll see the options:

- Link to Facebook, Twitter, and Google
- Manage my Contacts
- Manage my NOOK Friends
- Manage visibility of my LendMe books

All of these options are fairly self-explanatory. Manage my NOOK Friends gets a little more involved, however. Tap on that and you'll see the following categories:

- Friends
- Requests
- Sent

Friends will list all your current NOOK Friends—don't worry if you don't have any yet. All these friends will be shown in alphabetical order. Once you get friends, you can tap on a name to learn more about a friend's book collection or, if you hold your finger down, you have the option to delete the person from your friends list.

Requests lists any invitations you have from others to be their friend. For each request, you can tap Accept to become their friend or Reject to quietly reject their invite.

Sent shows people who have gotten your friendship invitations. If they are still on the list, then they haven't responded.

There are two ways to add new friends, and we'll talk about them in the following sections on Manually Inputting Contacts and Connecting to Facebook, Gmail, and Twitter.

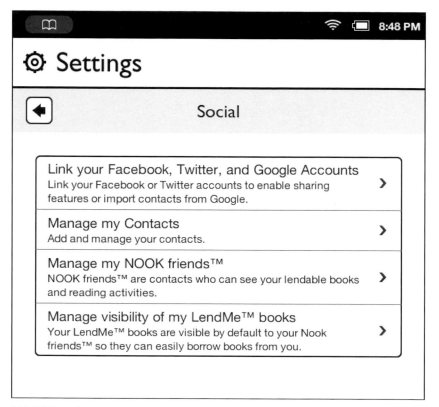

NOOK Social settings

LendMe

The LendMe Program lets you lend and borrow books from
friends or other people you're connected with on social networks.
Some books in the NOOK Store are lendable, and you'll be able
to borrow books, too. Any user of a NOOK, NOOK Color, NOOK
Tablet, or NOOK eReader App can lend and borrow books.

When you borrow a book, you will have it for 14 days. Afterward,
the book simply disappears from your device and is sent back to
the NOOK user who lent it to you. When a book is sent to a friend,
the recipient holds all of the digital rights to the book. In other

words, you can't access the book until it is returned returned 15 days later.

Keep in mind that not all books can be lent (that's up to the publisher), and those that can be lent can only be lent only once.

Manually Inputting Contacts

In order to lend a book, you first need a friend to lend it to. You can either manually input a person's name and email address (this only works if the person has a B&N account), or you can sync your NOOK with your Google account.

To connect with a person manually, go to the Social menu by pushing the Quick Nav button and tapping Social, then Manage my NOOK Friends. See the "+" sign? Touch it and type in their first name, last name, and email address. They will shortly receive an email about becoming your NOOK Friend.

Connecting to Facebook, Gmail, and Twitter

The NOOK can also connect to Facebook and Twitter. When connected to Facebook, you can use your NOOK to post updates about what books you're reading and learn more about what others are reading. Your NOOK doesn't monitor the interactions—it is just another way to share with others through your device.

The Twitter extension functions similarly. Once you're connected, NOOK can be used to tweet about whatever book you're reading, and will help foster an encouraging reading environment for you and your NOOK-owning friends.

Connecting to Google does something different from the other two networks. Google will let you swiftly scan your entire contacts list, looking for other people with NOOKs to use LendMe with. It will also import your Google contact list.

To use each of these features, push the Quick Nav button, and select Settings. From there, tap the Settings option labeled Social. In the Social menu, the first option should be to link your NOOK to Facebook, Twitter, and Google. Tap it, and there will be three buttons to link each respective account to your NOOK.

Lending a Book with LendMe

Now that you have NOOK Friends, you can let them borrow your stuff. The only books that can currently be lent are the works that were purchased from the NOOK Store, and that are in your Lifetime Library™. These books are yours forever and many of them can be lent to other people.

To lend a book:

> ▶ Open the book you'd like to lend.
> ▶ Tap the center of the page to get to your Reading Tools.
> ▶ Touch the More icon.
> ▶ Tap the Share button.
> ▶ Tap the LendMe icon.

Here, you may choose whether to lend a book to contacts on Facebook (by tapping the bubble next to each option). Tap Next, and navigate the Select a Friend menu to find the friend you want to send the book to. After you've found your friend, type in a message to them and tap Post. The book should be sent to the person, and he or she will have seven days to accept the book. If your friend doesn't accept, the book will simply be returned to you. If the friend does accept, the book will be returned 15 days after the person starts reading.

Using the NOOK Outside of the United States

NOOK is an American device that uses digital rights from publishers in the United States. However, you can still purchase content from the NOOK Store when you're outside the United States as long as you have a domestic billing address. You can also add other books to your NOOK by connecting it to a PC. Read more about this earlier in this chapter.

34.
Troubleshooting

D espite being a user-friendly device, NOOK can occasionally run into some technical problems. Here are a few fixes that might help with some potential problems.

Not Charging?

If your NOOK isn't charging, make sure the device is plugged in. You'll know it is being charged when the orange light at the bottom of the device is lit up. When the device is fully charged, the light turns green.

Not Connecting to Wi-Fi?

If you're having trouble connecting to Wi-Fi, make sure that your NOOK is connected and looking for the desired Wi-Fi network. Push the Quick Nav arrow press Settings and tap Wireless. From there, make sure that Wi-Fi is turned on and is connected. If it's not, either connect to an alternative Wi-Fi network, or troubleshoot your wireless router.

What Kind of Customer Support Does Barnes & Noble Offer?

Barnes & Noble understands that NOOK is an investment on your part, so the company has provided several ways for you to get the technical support you need:

- Phone
- Email
- Live online chat
- In-store

If you'd like to talk with a Barnes & Noble representative via your landline or cell phone, give the company a call at 1-800-843-2665.

If you prefer email, the Barnes & Noble troubleshooting email is nook@barnesandnoble.com.

You can also talk with someone online, which is like using an instant messenger program to ask an expert your pressing questions. To use the live online chat, go to http://www. barnesandnoble.com/nook/ support/ and click on the Chat Now link under Chat with a NOOK Expert. The Web site will ask you for five pieces of information:

- Name
- Email address
- Product (NOOK, NOOK Tablet, or NOOK Color)
- Serial or order number
- Question

Name is your full name and the Email Address is the one to which your Barnes & Noble account is registered. Be sure to choose the right product, as the troubleshooting for the NOOK, NOOK Color, and NOOK Tablet are different.

Also, it helps if you have your NOOK's serial number. If your NOOK is functional, simply press the NOOK button to open up the Quick Nav Bar, tap the Settings icon, touch Device Info, and then About Your NOOK. Here, you'll find lots of details about your device. The serial number will be the second-to-last number listed. It will be 16 digits.

Finally, describe your question as thoroughly as possible. Having a clear, detailed question will make the process faster and easier. The more info customer service has about the problem, the better they can help.

When you're ready, tap the Submit button. A Barnes & Noble representative will then hop online.

If you'd like to talk to someone face to face, you can go online at barnesandnoble.com and find a nearby Barnes & Noble store. The knowledgable associates will be happy to help.

Other Issues

If your problem is not too pressing, you can also use the Barnes & Noble forums. There's always an active community on the forums, so your question may have already been asked by someone else and answered by Barnes & Noble. You can check out the troubleshooting forums at http://bookclubs.barnesandnoble.com.

35. Accessories

Think you can't judge a book by its cover? Think again. NOOK users have a wide variety of options to protect and accessorize their devices, including cases, covers, bags, and even NOOK lights that will help your NOOK stand out.

Covers and Cases

Barnes & Noble sells an anti glare cover that can be applied on top of your NOOK's screen. This helps protect the device from things like dust and foreign particles while the screen still maintains all touch functionality. Your NOOK can also be styled with a variety of cases. These simply slide over your NOOK to add color, protection, and a look that is truly your own. The cases come in a variety of colors, designs, and materials. Even trendy designers like Jonathan Adler have created fun looks for NOOK.

Bags

In addition to cases, Barnes & Noble sells bags with pockets specifically designed to your NOOK. These cloth bags will keep your NOOK in a safe place, and also fit laptops up to fifteen inches.

Lights

Like most book lights, NOOK lights simply clip onto a convenient side of your NOOK to provide optimum reading light when you need it.

Other

Barnes & Noble also sells a stand for NOOK. This lets readers use the device, and keep it at an optimal reading angle without needing to hold it. You can also purchase car chargers for NOOK and replacement AC adapters.

APPENDICES

36. PubIt! and Publishing on NOOK

Do you have a book in you? Many people do, and everyone has a story to tell. It's a big challenge to find the time to write a book, and it's arguably an even bigger challenge to find a way to get a book into readers' hands. Publishing a print book is expensive, but cost-effective digital publishing has been much less successful than print...until now.

NOOK, NOOK Color, and NOOK Tablet have changed the game by making it easier than ever to get your book into an eager reader's hands through its PubIt! program.

Here are some PubIt! facts:

- It's free.
- It's automated online.
- It's a chance to make money.

Your book doesn't have to be a bestseller, either: Your grandma's recipes could be published for NOOK so your extended, tech-savvy family members can download them. Alternatively, if you think you have written something that other people will crave, publishing on NOOK gives you the opportunity to make a name for yourself. Heck, you may even make a little money.

Using PubIt!

The Barnes & Noble publishing program has four steps:

1. Write
2 Register
3. Upload
4. Publish

PubIt! Online

Write

Before you start thinking about publishing a book, you'll want to figure out what you'd like to write. There are literally thousands of books on writing books—or, more important, writing books well—so we won't get into that here. What we do recommend is spending some time thinking about your topic, spending some time writing it, and spending some time refining it.

To create the manuscript itself, you don't have to do anything special with the formatting. For instance, if you use Microsoft Word or Apple's Pages, feel free to use that format for your manuscript. More complex books need more complex formatting, but we're just going to stick with the basics for now.

Register

Next, head to the official PubIt! Web site at http://pubit. barnesand-noble.com. Here you can register for PubIt! using your Barnes & Noble email and password.

It will ask for the following info:

> ▌ First and last name
> ▌ Address
> ▌ Phone
> ▌ Publisher name and Web site

You are the publisher here, so feel free to put in your name and your Web site. This information will be listed with your book, so be sure to fill it out—you want to make sure that your fans can contact you! You'll also need a U.S. bank account and a taxpayer ID, which, if you don't own a company, is just your social security number. Unfortunately, as of December 2011, PubIt! only supports United States publishers.

The final setup step is giving your bank information with your routing number, checking account, taxpayer ID, and company details. PubIt! uses a direct-deposit system, so that's how you'll get paid. Click Submit, and you are officially a NOOK publisher.

Upload

Now you can upload your manuscript. There are several details you need to fill in for your book, including:

> ▌ Title
> ▌ Sale price

- Cover
- Search data

Let's go over the most important parts. First, you want to make sure that the title fits the topic as well as the audience you're going after. Naming your book *Chainsaw Murder 3* and gearing it toward nice retirees isn't the best plan. Check out other books similar to yours and see which titles stand out.

Sale price is totally up to you and is one of your more crucial decisions. Many books on PubIt! earn 65 percent of their cover price, so if you make your book price $5.00, each sale will net you $3.25. That's actually not bad at all, especially since traditional publishing usually gives authors about 10 percent of the cover price. Punch in a price and, using the link next to the royalty box, PubIt! will tell you what your pay will be. Again, look at other books similar to yours. If a similar book costs $.99, you probably shouldn't charge $5.99 for yours.

Producing a cover is perhaps the most difficult part since, if you're like most of us, you aren't necessarily artistically talented enough to design your own. Do a quick search online and you'll find lots of low-cost cover artists, though you'll want to ask them for references and examples of their work. An even better place to find a cover creator is on the Barnes & Noble forums. However, before you contact any artist, think about what you'd like the cover to convey. If your book project is just a quick experiment for you, or it's only going to be passed around to your family and friends, feel free to knock out a cover yourself.

Finally, the search details will help your book get noticed. There's lots to fill in here, including:

> ▶ Product description
> ▶ Reviews from other people
> ▶ Type of book

We recommend filling in as much as you can. Think about the information you notice when you're looking for books in the NOOK Store yourself: You're reading what other people thought of it, looking at the cover, getting a feel from the description, and so on.

Publish

When you are ready, Barnes & Noble will publish your book on PubIt!. Download the book immediately—well, buy it from yourself—and check for errors. You can always correct the original manuscript, re-upload the book, and have the new version available within minutes.

It is also time to promote your book. Use your social network connections on Twitter, Facebook, LinkedIn, or others. Let everyone know you have a book available. Most importantly, let them know that you are open to feedback: If they don't like the book, ask them to tell you why; if they love it, ask them to tell everyone about it.

37. Online Resources

There are many sources online to help you get the most out of your NOOK. There are also several different sources for free, public-domain eBooks.

Barnes & Noble
bn.com

Naturally, bn.com is the first site on the list. Here, you can browse and download NOOK books from your computer, smartphone, or tablet. You can also purchase real books, add more credit cards to your BN account, and purchase other things like accessories for your NOOK. You can also buy a NOOK here if you don't have one yet.

BN Book Clubs
bookclubs.barnesandnoble.com

Want to talk about books with like-minded NOOK owners? Barnes & Noble has you covered. Here, you'll be able to find announcements, discussions, and other news from the NOOK world.

Gmail, Twitter, Facebook

NOOK's social functions are amplified by being able to look for NOOK Friends though other social networks. Having Gmail, Twitter, and Facebook accounts greatly increases the number of people you can share books with. Another way to connect with fellow readers is at https://www.facebook.com/nookBN.

ProjectGutenberg.net, ManyBooks.net

These sites are free sources for copyright-free eBooks and texts. All of the books are available at no cost and can be a great way to add quality content to your NOOK Library.

38. Glossary

Charger · Wall adapter that plugs into NOOK to recharge the battery.

Daily Shelf · A row of books, magazines, and newspapers running on the bottom of the Home Screen for NOOK Color and NOOK Tablet, displaying items that have been recently downloaded.

Device Info · Settings screen detailing NOOK's software.

eBook · A digital book read on an eReader.

Home · NOOK's front screen.

Ion Lithium Battery · A rechargeable battery that doesn't need replacing.

LendMe · Barnes & Noble's program to let friends borrow books.

Library · The list of books you own or have borrowed.

NOOK · The Barnes & Noble family of eReaders.

Quick Nav Button · The button near the bottom of your NOOK. Use it to navigate NOOK's menus quickly.

Screensaver · Default screen on NOOK when not in use.

Search · The option to look for a book based on keywords, as well as search the NOOK Store and also (on NOOK Color and NOOK Tablet) to look for music and other downloadable items.

Settings · A group of screens with different configuration settings.

Store · The Barnes & Noble online store, where you can purchase new books and samples.

Side Buttons · The button on the side of the NOOK that turns the page.

Social · The menu for social settings.

Swipe · A move from left to right, or right to left, with a finger on the screen.

Tap · A simple press on the screen with your finger.

Touchscreen · The screen on your NOOK that responds to being touched.

USB · Plug that connects to nearly every computer.

Wi-Fi · Wireless Internet connection.

Wireless · The page detailing Wi-Fi information.

39. FAQ (Frequently Asked Questions)

My NOOK was on, but now it has a blank screen. How do I turn it back on?

Leave your NOOK unattended and it will go into "sleep mode," which means it is still on, but it has turned off the screen so you won't waste any battery life. To turn your NOOK back on, tap the "n" button at the bottom center of the device. The screen will now "wake up." See the "n" symbol on the screen? Touch it, put your finger down firmly, and slide the symbol from the left to the right of the screen. Your NOOK is now awake.

My NOOK was working fine, but it suddenly froze. How can I get it working again?

It won't happen often, but like any computer, your NOOK can sometimes freeze up while you're using it. Most of the time, it is nothing to worry about. All you need to do is a "soft reset." To do a soft reset, you first need to locate the Power button. On NOOK, the power button is at the very top of the device. On NOOK Color and NOOK Tablet, the Power button is on the left side of the device near the top. Once you find it, touch and hold down the Power button for 20 seconds. Your NOOK will power down. Now touch and hold the power button for a couple seconds again. It should power back up and bring you back to where the device originally froze.

My NOOK is responding to button presses, but I'm not touching it! What should I do?

As with any touchscreen device, sometimes dirt or other things will interfere with how receptive your NOOK is to your touch. In fact, in some cases, it can actually make NOOK think you are touching it when you're not. It's probably not a major issue here—just a matter of cleaning your NOOK touchscreen. To do that, get a lightly damp, soft cloth. We highly recommend a microfiber cloth or another material specifically made for wiping down sensitive surfaces, like computer screens or eyeglass lenses. Turn off your NOOK and, once it is completely powered down and unplugged from your electrical outlet and computer, give the touchscreen a thorough, gentle cleaning. Give it time to dry and, once the NOOK screen is no longer damp, turn it on. It should be ready to go.

I want to give my NOOK to someone else, but I don't want all my private documents on there. How do I "clean" my NOOK?

What you need to do is de-register your NOOK through the Settings menu. On NOOK, tap the "n" button to get to the Options menu and select Settings. On NOOK Color and NOOK Tablet, pull up the Quick Nav menu by tapping the "n" button, tap the Settings icon and, finally, tap the Device Info arrow. When you're ready to clean your device, tap the Erase and Deregister Device option. It will now remove all the documents, books, music, or other items on your NOOK, as well as the link to your registration from the device. The books you purchased won't be on your NOOK anymore, but they will be securely stored in the Barnes & Noble Archive, so you can re-download them free of charge whenever you get another NOOK.

I'm not sure if I should upgrade my warranty or even if my NOOK is still covered. How do the basic and advance warranties work?

The basic warranty is included with any NOOK; it protects your device from malfunction and provides free customer service for one year. The advanced B&N Protection Plan costs additional money but will protect your device after accidents like spills and cracks, and it extends your protection to a total of two years. If you return a malfunctioning NOOK, Barnes & Noble will evaluate the case and, depending on your warranty and the damage done, will either fix and return your NOOK or send you a refurbished, functional NOOK. If the problem you're having is not covered under your warranty, or if your warranty has expired, Barnes & Noble will reject the claim and just send you back the device unchanged.

Index

The Third COAST